I0054227

Contents

Chapter One

Introduction

A few years ago, I stood on the balcony of my tiny urban apartment, staring at a row of sad, wilting plants. I dreamed of lush green vegetables and vibrant flowers, but the reality differed. The soil in my containers was dry and lifeless, and my plants struggled to survive. Frustrated, I realized I needed to understand what was happening beneath the surface. That is when I discovered the transformative power of soil science.

This book is born out of that journey. Its purpose is simple: to provide practical, relatable steps for managing soil in urban gardening. Whether working with containers, raised beds, or any limited space, this guide aims to help you grow healthy, thriving plants.

Why focus on soil science? Because it is the foundation of any successful garden. Healthy soil means healthy plants. Did you know that a teaspoon of good soil contains more microorganisms than people on Earth? These tiny organisms play a vital role in breaking down organic matter and releasing nutrients that plants need. Soil health directly affects plant health and understanding it can make all the difference in your urban garden.

This book is for urban gardeners like you who are interested in sustainable practices and soil health. It is for those who live in North America, Europe, Africa, the Middle East, and Australia. You value eco-friendly gardening and want to maximize your limited space.

The structure of this book is straightforward. First, we explain the basics of soil science—what soil is, how it works, and why it matters. Then, we address practical steps to improve soil quality in containers and raised beds. We will cover composting, organic pest control, and water management. Finally, we will explore advanced techniques and sustainable practices to keep your garden productive year-round.

What can you expect to gain from this book? Practical benefits, for starters. You will learn to maximize your limited space, whether it is a balcony, rooftop, or small backyard. You will learn how to improve soil quality, organically manage pests, and maintain a productive garden throughout the seasons. These are actionable steps you can take today.

Let me share a few success stories to illustrate the power of good soil management. In Chicago, a rooftop gardener transformed her barren space into a thriving vegetable garden by focusing on soil health. In Berlin, a community garden project turned an empty lot into a green oasis, feeding dozens of families with fresh produce. These stories, and others like them, show what is possible when we understand and care for our soil.

My vision for this book is to empower you with the knowledge and tools to create optimal raised bed gardens. You will grow nutritious vegetables no matter what the climate. Whether you are a beginner or an experienced gardener, this book offers something for everyone.

I invite you to engage with the content, experiment with the techniques, and share your gardening successes. Think of this book as a hands-on guide you can refer to throughout your gardening journey.

Together, we can unlock the secrets of soil science and create thriving urban gardens.

So, let us get started. The journey to a healthier, more productive garden begins beneath the surface, in the rich, living world of soil.

Chapter Two

Urban Gardening and Soil Science

When I first moved to the city, having a garden seemed like a distant dream. I had grown up surrounded by sprawling gardens and lush greenery, but now, my reality was a small apartment with a tiny balcony. Determined to bring some of that green into my urban life, I started planting in containers. However, my initial attempts could have been better. The plants did not thrive, and I could not figure out why. It was when I understood the importance of soil that everything changed. This book is about sharing that revelation to you and helping you turn your urban space into a thriving garden.

The book provides practical and relatable steps for managing soil in urban gardening. Whether growing plants in containers, raised beds, or any limited space, this guide aims to help you succeed. By focusing on soil science, you can overcome many challenges urban gardeners face and achieve a healthy, productive garden.

Understanding soil science is crucial for urban gardeners. It is easy to overlook the importance of soil when you are focused on planting and watering seeds. However, your soil's health directly affects your plants' health. For instance, did you know that one gram of healthy soil can contain up to a billion bacteria, a million fungi, and other microorganisms? These organisms play a vital role in breaking down organic matter and releasing nutrients that plants need. Without healthy soil, even the most diligent watering and fertilizing will not yield the results you want.

Urban gardeners interested in sustainable practices and soil health will find this book tailored to their needs. It is written with a global perspective, relevant to readers in North America, Europe, Africa, the Middle East, and Australia. You likely value eco-friendly practices and strive to maximize your limited space for growing food or enhancing your environment.

We will start by exploring the basics of soil science, including what soil is, how it works, and why it matters. Then, we will take practical steps to improve soil quality in containers and raised beds. We will cover composting, organic pest control, and water management topics. Finally, we will explore advanced techniques and sustainable practices to keep your garden productive year-round.

By the end of this book, you will gain practical benefits. You will learn to maximize limited space, whether a balcony, rooftop, or small backyard. You will learn how to improve soil quality, organically manage pests, and maintain a productive garden through the seasons. These are actionable steps you can take today to see immediate improvements.

Throughout the book, I will share success stories that illustrate the transformative power of good soil management. For example, a rooftop garden in Chicago was turned into a thriving vegetable oasis by focusing on soil health. A community garden project in Berlin transformed an empty lot into a green sanctuary, providing fresh produce for dozens of families. These stories show what is possible when we understand and care for our soil.

My vision for this book is to empower you with the knowledge and tools to create optimal raised bed gardens. You will grow nutritious vegetables no matter what the climate. Whether you are a beginner or an experienced gardener, this book offers something for everyone.

Image showing a vibrant raised bed garden

I encourage you to engage with the content, experiment with the techniques, and share your gardening successes. Think of this book as a hands-on guide you can refer to throughout your gardening endeavors. Together, we can unlock the secrets of soil science and create thriving urban gardens.

2.1 Understanding Soil Science for Urban Environments

Soil science might sound complex, but at its core, it is about understanding the medium in which plants grow. Soil is a dynamic, living entity comprising minerals, organic matter, water, and air. These components interact in many ways to support plant life. Minerals provide the essential nutrients plants need. Organic matter, like decomposed leaves and compost, improves soil structure and fertility. Water is necessary for nutrient transport to plant roots, while soil air enables root respiration and supports beneficial microorganisms.

Image showing soil horizons

In natural settings, soil forms distinct layers called horizons. The top layer, known as the O horizon, is rich in organic material. Below that lies the A horizon, or topsoil, where most root activity occurs. The B horizon, or subsoil, contains minerals leached from the upper layers, and the C horizon consists of weathered parent material. Understanding these layers can help you manage your soil more effectively, even in an urban environment.

Urban gardening comes with unique challenges, especially regarding soil. One major issue is soil compaction. Foot traffic, construction, and heavy machinery can compress soil particles, reducing air pockets and making it difficult for roots to penetrate. Compacted soil also hampers water infiltration, leading to poor drainage and root oxygenation. Urban soils often face contamination from heavy metals like lead and chemicals from industrial activities. These contaminants can pose health risks and affect plant growth.

Different soil types have distinct properties that can influence gardening success. Sandy soil drains quickly but does not hold nutrients well. Clay soil retains water but can become compact and hard. Silt soil has a smooth texture and holds nutrients but is prone to erosion. Loam soil, a balanced mix of sand, silt, and clay, is ideal for most plants because of its excellent drainage and nutrient-holding capacity. Understanding your soil's texture and structure can help you amend it to suit your gardening needs better.

Regular soil testing is essential to understand soil health and composition. Testing for pH levels can tell you whether your soil is too acidic or alkaline for optimal plant growth. Most plants prefer slightly acidic to neutral pH. Soil tests also reveal nutrient availability, helping you identify deficiencies or excesses that need correction. Testing can also detect contaminants, allowing you to take necessary precautions to ensure a safe growing environment. By regularly testing your soil, you can make informed decisions to improve and maintain its health, leading to a more productive and thriving urban garden.

2.2 The Importance of Soil Health in Urban Gardening

Healthy soil is the cornerstone of any thriving garden, especially in urban settings with limited space and resources. Soil health refers to the soil's ability to sustain plant and animal productivity, maintain environmental quality, and promote plant and animal health. A complex interplay of factors, including soil biodiversity and organic matter, creates a robust ecosystem. Soil biodiversity encompasses a variety of life within the soil, such as bacteria, fungi, earthworms, and other microorganisms. These organisms break down organic material, fix nitrogen, and decompose plant residue, contributing to the overall fertility of the soil. Organic matter, such as decomposed plants and animal residues, serves as a reservoir of nutrients and improves soil structure, enhancing its capacity to retain water and support plant roots.

There are many significant benefits to maintaining healthy soil. One of the primary advantages is enhanced nutrient uptake. Healthy soil provides a balanced supply of essential nutrients, allowing plants to absorb what they need for growth and development. It means your plants will be stronger, more resilient, and yield better. Improved water retention is another crucial benefit. Healthy soil with good organic matter can hold water more effectively, reducing the need for frequent watering and helping plants survive during dry spells. Healthy soil helps reduce erosion by binding soil particles together, essential in urban areas where soil can be easily disturbed.

Identifying healthy soil involves observing several vital indicators. Soil texture and structure are fundamental aspects to consider. Healthy soil typically has a crumbly texture, neither too sandy nor too clayey, allowing pleasant air and water movement. The presence of earthworms and microorganisms is another positive sign. Earthworms aerate the soil and help decompose organic material, while a diverse microbial population shows a balanced and fertile soil. Digging into your soil and finding it teeming with life shows it is healthy.

You can take several practical steps to improve and maintain soil health in your urban garden. Adding compost and organic matter is one of the most effective methods. Compost enriches the soil with nutrients, improves its structure, and boosts its water-holding capacity. You can make your compost using kitchen scraps, garden waste, and other organic materials or purchase it from a trusted supplier. Another beneficial practice is cover cropping, which involves planting specific crops, such as clover or rye, which help protect and enrich the soil during the off-season. Cover crops prevent erosion, improve soil structure, and add organic matter when decomposing. Mulching is also a valuable technique. Covering the soil with organic materials like straw, wood chips, or leaves can reduce moisture loss, suppress weeds, and contribute to the organic matter as the mulch breaks down.

Soil Health Checklist

- Soil Texture: Crumbly, well-aggregated soil that feels good in your hand.

- Earthworms and Microorganisms: Look for earthworms and a variety of tiny organisms.

- Organic Matter: Ensure a good mix of decomposed plant and animal material.

- Water Retention: Soil should hold moisture without becoming waterlogged.

- Nutrient Balance: Regularly test for essential nutrients and adjust as needed.

Improving soil health in urban environments can be a significant change for any gardener. For example, a city gardener in Toronto once struggled with dry, compacted soil in her small backyard. She transformed her garden into a lush, productive space by starting a composting routine and incorporating more organic matter into her soil. The plants grew more robust and healthier; she noticed they required less frequent watering.

Remember, healthy soil is about growing plants and creating a sustainable ecosystem in your garden. Each step you take to improve your soil contributes to a larger goal of environmental sustainability. Whether you add compost, plant cover crops, or observe the life in your soil, you are playing a part in a much bigger picture. Gardening has become not just a hobby but a meaningful way to connect with nature and contribute to the health of our planet.

2.3 Overview of Urban Gardening: Containers, Raised Beds, and Limited Spaces

Urban gardening offers several methods to cultivate plants in confined spaces, each with distinct advantages and challenges. Container gardening involves using pots, planters, and vertical gardens. This method is highly adaptable and perfect for balconies, rooftops, and windowsills. Container gardening allows you to move plants to optimize sunlight and protect them from adverse weather. However, containers can dry out quickly and may require frequent watering. Raised bed gardening, on the other hand, involves constructing materials like wood, metal, or recycled items. Raised beds offer greater control over soil quality and drainage. They make gardening accessible to those with mobility issues and can be placed on any surface, from concrete patios to significant challenges. The major challenge is the initial setup, which can be labor-intensive and costly. Creative use of small spaces is another approach. This practice includes transforming balconies, rooftops, and windowsills into green havens. Small spaces require clever use of vertical and horizontal space, often combining container and raised bed gardening elements.

Each method has its pros and cons. Container gardening offers unmatched mobility. You can move your plants to chase the sunlight or bring them indoors during a storm. It is also space-efficient, allowing you to grow various plants in a limited area. However, containers can be prone to drying out, requiring more frequent watering and attention to soil health. Raised beds provide excellent soil control. You can tailor the soil mix to your plants' needs, ensuring optimal growth conditions. Raised beds also improve accessibility, making it easier to tend to your garden without bending or kneeling. The downside is the initial effort and expense involved in setting them up. When using small spaces, the creative use of every inch is crucial. Vertical gardens, hanging baskets, and tiered planters can maximize space. The challenge lies in ensuring that all plants receive adequate sunlight and water, which may require innovative solutions like drip irrigation or self-watering systems.

Your garden can make a big difference in how well your plants grow. For container gardening, choose the right soil mix. Compost, peat moss, and perlite blend work well for most plants. Ensure good drainage by placing gravel at the bottom of the container or using pots with drainage holes. Watering is crucial as containers dry out faster than ground soil. Check

Image showing compost, perlite blend and peat moss

the soil moisture regularly and water deeply but less frequently to encourage root growth. Raised beds benefit from a well-prepared soil mix, too. Combine native soil with compost and organic matter to enrich the soil. Mulch the surface to retain moisture and suppress weeds. Install a drip irrigation system to ensure consistent watering. For small spaces, select plants based on their light and space requirements. Herbs, like basil and thyme, thrive in small pots while climbing plants like tomatoes and beans can be trained on trellises. Use reflective surfaces to maximize light and consider portable planters to move plants as needed.

Global examples of successful urban gardens offer inspiration and practical guidance. A cramped balcony in New York City was ingeniously converted into a verdant garden, using vertical structures and smart container choices to cultivate various herbs, flowers, and vegetables within a confined space. Similarly, a deserted rooftop in Tokyo was reimagined into a thriving sanctuary, employing raised beds crafted from repurposed materials and a blend of local soil and compost to achieve superior soil management and moisture regulation. In Berlin, a community garden initiative revitalized an abandoned plot, integrating raised beds and container gardening to foster a diverse plant ecosystem. This endeavor maximized the utility of limited urban spaces, strengthened community bonds, and underscored the importance of ecological responsibility.

Urban gardening is about making the most of what you have - a tiny balcony, a rooftop, or a small backyard. Each method—container gardening, raised beds, and creative use of small space offers unique benefits and challenges. If you follow these simple steps and use the proper methods, you can have a lovely garden even in the city. From choosing the right soil mix to ensuring proper watering, every step brings you closer to a thriving urban garden. The success stories from New York City, Tokyo, and Berlin show that anyone can turn limited space into a green haven with creativity and effort.

2.4 Soil Contamination in Urban Areas: Identification and Solutions

Urban gardening often involves soil contamination, a significant challenge for many city dwellers. Common soil contaminants like heavy metals—lead, mercury, and cadmium—pose severe risks to human health and plant growth. Lead, particularly prevalent because of its historical use in gasoline and paint, can linger in the soil for decades. Mercury and cadmium, often byproducts of industrial activities, further complicate the issue. Chemical pollutants, including pesticides and industrial waste, add another layer of complexity. These contaminants can enter the soil through various pathways, such as air deposition, runoff, or improper disposal, making urban soil a potential hazard zone. Understanding these contaminants is the first step toward creating a safe and productive garden.

Testing soil for contaminants is crucial before planting, as it helps identify potential risks and guides remediation efforts. Home testing kits offer a convenient way to get an initial assessment. These kits typically test for common contaminants like lead and other heavy metals, pH levels, and nutrient content. They are user-friendly and provide quick results, making them ideal for preliminary testing. However, professional soil analysis services are recommended for a more comprehensive analysis. These services use advanced techniques to test various contaminants, including industrial chemicals and pesticides. The Environmental Protection Agency (EPA) recommends testing for pH, organic matter, nutrients, micronutrients, metals, and lead. Professional testing provides detailed insights crucial for effective remediation and long-term soil health.

Once contaminants are identified, various remediation techniques can make the soil safe for gardening. Phytoremediation, a method that uses plants to absorb contaminants, is an eco-friendly solution. Certain plants, such as sunflowers and mustard greens, absorb heavy metals from the soil. These plants can be grown and safely disposed of, gradually reducing contamination. Soil replacement is another effective method, especially for highly contaminated areas. This practice involves removing and replacing the contaminated soil with clean, uncontaminated soil. Barrier methods, such as geotextiles or other physical barriers, can prevent contaminants from migrating into the gardening area. Place these barriers under raised beds or the garden perimeter to create a safe growing environment.

Preventing future contamination is equally essential to maintain a healthy urban garden. Using raised beds filled with clean soil is one of the most effective strategies. Raised beds provide better control over soil quality and reduce the risk of contaminants from the surrounding area leaching into the garden. Regular soil testing and monitoring are also crucial. By keeping track of soil health and contamination levels, gardeners can take timely action to address any issues. Implementing good gardening practices, such as using organic fertilizers and avoiding chemical pesticides, can further minimize the risk of future contamination. Engaging in community efforts to clean up and monitor local soil can contribute to broader environmental health.

City Slicker Farms in Oakland, California, is a practical example of effective soil remediation. This urban farm was established on a former industrial site heavily contaminated with lead and other pollutants. The farm used soil replacement, phytoremediation, and raised beds to create a safe and growing environment. Community involvement was crucial, with volunteers participating in soil testing, planting, and remediation efforts. Today, City Slicker Farms stands as a testament to the power of collective action and sustainable practices in transforming urban spaces.

Understanding and addressing soil contamination requires a multifaceted approach. Testing is the first critical step, providing the information needed to choose the proper remediation techniques. Whether through phytoremediation, soil replacement, or barrier methods, there are practical solutions to make urban soil safe for gardening. Preventing future contamination involves ongoing vigilance and adopting sustainable practices. These steps allow urban gardeners to create secure, productive gardens contributing to personal and community well-being.

2.5 The Role of Microclimates in Urban Gardening

Microclimates are small-scale climate zones within a larger area and play a crucial role in urban gardening. In cities, temperature, humidity, and sunlight variations can be significant, even within a few feet. Buildings, roads, and other infrastructure influence these variations, creating unique growing conditions. For instance, a south-facing wall might absorb heat and create a warm microclimate, while a shaded courtyard might stay cool and moist. Understanding these microclimates can help you choose the right plants and optimize your gardening practices.

Identifying your specific microclimate involves careful observation and measurement. Start by observing sun patterns in your garden. Note which areas receive full sun, partial or complete shade throughout the day. Pay attention to seasonal changes as well. In summer, the sun is higher and casts different shadows compared to winter. Next, measure temperature and humidity variations. A simple thermometer and hygrometer can provide valuable data. To comprehensively understand your garden's microclimate, record these measurements at various times of the day and weather. Consider the influence of surrounding structures. Buildings can block wind, create shade, and reflect heat, all affecting your garden's microclimate.

Adapting your gardening practices to suit these microclimates is critical to success. Start by selecting appropriate plants. For example, choose shade-tolerant plants for areas that receive limited sunlight.

Ferns, hostas, and some herbs like mint thrive in shady conditions. In contrast, drought-resistant plants like succulents and lavender are ideal for hot, sunny spots. Adjust your watering and fertilization schedules based on microclimate needs. Shaded areas may require less frequent watering, while sunny spots might need more. Similarly, fertilization should be tailored to plants' specific needs in different microclimates. Keep a gardening journal to track what works and what does not, adjusting as necessary.

Real-world examples illustrate how gardeners successfully adapt to their microclimates. On a north-facing balcony in Melbourne, a gardener found that most plants struggled with the lack of direct sunlight. She transformed the space into a lush retreat by switching to shade-tolerant plants like ferns and hostas. In contrast, a rooftop garden in Dubai faced intense heat and direct sunlight. The gardener chose drought-resistant plants like succulents, cacti, and Mediterranean herbs, which thrived despite the harsh conditions. He further optimized the garden's microclimate by using reflective materials and installing a drip irrigation system, ensuring consistent growth and minimal water waste.

Another example is a community garden in Berlin, which had varied microclimates because of the surrounding buildings. Some plots received full sun, while others were mostly shaded. Gardeners selected plants based on these conditions and adjusted their care routines accordingly. Sun-loving vegetables like tomatoes and peppers thrived in the sunny plots, while leafy greens and herbs flourished in the shaded areas. By understanding and adapting to their microclimates, these gardeners maximized productivity and created a diverse, vibrant garden.

Microclimates are a decisive factor in urban gardening. You can create optimal plant growth conditions by understanding and leveraging these small-scale climate zones. Start by observing and measuring your garden's unique conditions. Choose plants well-suited to these microclimates and adjust your care routines as needed. Real-world examples show that you can turn challenging situations into opportunities with the right approach. Adapting to your microclimate can lead to a thriving urban garden, whether you are dealing with shade, heat, or wind.

As we conclude this exploration of microclimates, remember that every urban garden is unique. Many factors influence your microclimate, from the orientation of your home to the materials used in your garden structures. By taking the time to understand and adapt to these conditions, you can create a successful and sustainable garden. Embrace the diversity of your urban environment and use it to your advantage. Your garden will thank you with lush growth, vibrant colors, and bountiful harvests.

Chapter Three

Container Gardening Essentials

W hen I moved to a bustling city, I missed the sprawling garden I had grown up tending. Determined to bring green into my urban space, I turned to container gardening. My first attempts were a mix of successes and failures, but they taught me valuable lessons about choosing the proper containers. The correct container can make all the difference in the health and productivity of your plants. This chapter will guide you through selecting the best containers for your urban garden, ensuring you make choices that benefit your plants and living space.

4.1 Selecting the Right Containers for Your Urban Garden

Choosing the proper container can be a transformative step for your urban garden. Different containers offer various advantages and are suitable for other plants. Clay pots, for example, are excellent for aeration. They allow air to permeate the soil, which benefits root health. However, they are heavy and fragile, making them less ideal for larger plants or if you need to move your containers frequently. Plastic pots, on the other hand, are lightweight and retain moisture well. They are perfect for those who may need to remember to water regularly, as they reduce the frequency of watering required. Fabric grow bags are another innovative option. They promote air pruning of roots, which prevents plants from becoming root-bound and encourages root growth. Last, self-watering containers are a godsend for busy gardeners. These containers have a built-in reservoir that supplies water to the plant roots, ensuring consistent moisture levels and reducing the need for frequent watering.

The size and depth of your containers are crucial factors to consider, as different plants have different root systems and space requirements. Herbs and small plants, like basil and parsley, do well in shallow containers. These plants have relatively small root systems that don't require deep soil. Root vegetables, such as carrots and radishes, need deeper containers to allow their roots to develop fully. A container at least 12 inches deep is ideal for these plants. Large plants, like tomatoes and peppers, require robust, broad containers. These plants need ample space for their roots and the stability a larger container provides to support their growth and fruit production.

Proper drainage is essential for container gardening. Without it, excess water can accumulate at the bottom of the container, leading to root rot and other issues. Most containers have drainage holes, but you can easily add them using a drill if yours does not. Aim for several holes to ensure adequate drainage. To further improve drainage, you can add a layer of potting gravel or broken pottery at the bottom of the container. This creates a space for excess water to collect away from the plant roots, reducing the risk of waterlogging.

Balancing aesthetic appeal with functionality is essential when selecting containers. You want your garden to be visually pleasing but also practical. Matching containers to your decor can create a cohesive look, making your garden an extension of your living space. Consider the style and color of your containers and how they complement your overall aesthetic. Portability is another factor to consider. Lightweight containers, such as plastic pots and fabric grow bags, are easier to move around, allowing you to reposition plants to optimize sunlight exposure. Sustainability is increasingly important to many urban gardeners. Choosing eco-friendly materials, like biodegradable pots or containers made from recycled materials, aligns with sustainable gardening practices and reduces your environmental footprint.

Container Selection Checklist

Container Type

- Clay Pots: Good for aeration but heavy and fragile.

- Plastic Pots: Lightweight and retain moisture well.

- Fabric Grow Bags: Promote air pruning of roots.

- Self-Watering Containers: Ideal for busy gardeners.

Size and Depth

- Herbs and Small Plants: Shallow containers.

- Root Vegetables: Deep containers.

- Large Plants: Robust, broad containers.

Drainage
- Add Drainage Holes: Ensure proper drainage.

- Use Potting Gravel: Improve water flow.

Aesthetic and Functional Considerations
- Match Decor: Create a cohesive look.

- Portability: Choose lightweight options.

- Sustainability: Opt for eco-friendly materials.

Selecting the correct container is foundational to successful urban gardening. It affects everything from plant health to garden aesthetics. By considering your containers' type, size, depth, drainage, and aesthetics, you can create an environment where your plants will thrive. Each choice you make should cater to the specific needs of your plants while also fitting seamlessly into your living space. This thoughtful approach ensures that your urban garden is productive and beautiful, providing a green oasis in the city's heart.

4.2 Soil Mixes for Containers: Recipes and Tips

Creating the perfect soil mix for your containers is crucial for the success of your urban garden. A well-balanced blend provides nutrients, retains moisture, and ensures proper aeration and drainage. The essential components of a good soil mix include compost, peat moss or coconut coir, and perlite or vermiculite. Compost is nutrient-rich organic matter that feeds your plants and improves soil structure. It's the backbone of any good mix, providing a steady supply of nutrients as it breaks down. Peat moss or coconut coir is essential for moisture retention. These materials hold water well, preventing the soil from drying out too quickly, which is especially important in containers that can lose moisture faster than ground soil. Perlite or vermiculite is included to enhance aeration and drainage. These lightweight materials keep the soil from compacting and allow excess water to drain away, reducing the risk of root rot.

Different plants have different needs, and customizing your soil mix can significantly affect their health and productivity. A simple blend of equal parts compost, peat moss, and perlite works well for herbs. This combination provides the right balance of nutrients, moisture retention, and aeration, ensuring your herbs have everything they need to thrive. Succulents and cacti, on the other hand, require a mix that is well-draining and low in organic matter. A good recipe for these plants is to use more sand and less compost—about 60% sand, 30% compost, and 10% perlite. This gritty mix mimics the natural environment of succulents, allowing water to drain quickly and preventing root rot. Acid-loving plants like blueberries and azaleas benefit from a blend with additional peat moss and pine bark. The peat moss lowers the pH, making the soil more acidic, while the pine bark provides structure and aeration.

Maintaining soil health in containers requires regular attention and care. One of the simplest ways to keep your soil healthy is by regularly adding organic matter. Compost is an excellent choice, as it continuously replenishes nutrients and improves soil structure. Adding a layer of compost to the top of your containers every few months can make a big difference. Flushing out salt from fertilizers is another essential practice. Over time, salt can build up in the soil from fertilizers, affecting plant health. To flush them out, water your containers thoroughly until water drains from the bottom, carrying away excess salts. This practice should be done periodically, mainly if you use synthetic fertilizers.

Reusing and refreshing container soil is a sustainable practice that saves time and money. Remove old roots and debris from the soil at the end of each growing season. This practice prevents disease and pests from overwintering in your containers. Next, fresh compost and other amendments should be mixed to replenish nutrients and improve soil structure. If the soil has become compacted or depleted, consider adding perlite or vermiculite to improve aeration and drainage. By refreshing your soil, you ensure your plants start the new season with a healthy growing medium.

Container Soil Maintenance Checklist

Components of Ideal Soil Mixes

- Compost: Nutrient-rich organic matter.

- Peat Moss or Coconut Coir: For moisture retention.

- Perlite or Vermiculite: For aeration and drainage.

Customizing Soil Mixes for Different Plants
- Herb Mix: Equal parts compost, peat moss, and perlite.

- Succulent Mix: More sand and less organic matter.

- Mixed Acid-Loving Plant: Additional peat moss and pine bark.

Tips for Maintaining Soil Health
- Regularly Adding Organic Matter: Top with compost every few months.

- Flushing Out Salts: Thoroughly water to remove excess salts.

Reusing and Refreshing Container Soil
- Removing Old Roots and Debris: Prevent disease and pests.

- Mixing in Fresh Compost and Amendments: Replenish nutrients and improve structure.

Creating and maintaining the right soil mix for your containers is an ongoing process that pays off in healthier, more productive plants. By understanding your plants' needs and providing a well-balanced soil mix, you can create an ideal environment for growth. Regular maintenance, including adding organic matter and refreshing soil, ensures your containers remain fertile and well-structured season after season. Whether you grow herbs, succulents, or acid-loving plants, the right soil mix is critical to your success.

4.3 Watering Techniques for Container Gardens

Understanding the watering needs of your plants is fundamental to successful container gardening. Several factors affect how much water your plants will need. Plant type is a primary consideration. Succulents and cacti, for instance, need far less water than leafy greens like spinach or herbs like basil. Container size also plays a role; smaller containers dry out faster and may require more frequent watering. Weather conditions are another crucial factor. During hot, dry spells, your plants will need more water, whereas cooler, humid conditions may reduce their water requirements. Over-watering and under-watering both present challenges. Signs of over-watering include yellowing leaves and a soggy soil surface. Under-watering often results in wilting, dry soil, and brown leaf edges. Monitoring these indicators will help you adjust your watering practices accordingly.

Efficient watering methods can save you time and water while ensuring your plants get the moisture they need. Drip irrigation systems are effective. These systems deliver water directly to the plant roots through tubes and emitters, providing consistent humidity and reducing water waste. They are ideal for busy gardeners needing more time for frequent hand watering. Self-watering containers are another excellent option. These containers feature a reservoir that supplies water to the plant roots as needed. This setup reduces the watering frequency and helps maintain a consistent moisture level, which is especially beneficial for plants that prefer even soil moisture. Hand watering remains a popular method for many gardeners. It allows you to control the amount of water each plant receives and ensures even moisture distribution. Using a watering can with a fine rose or a gentle spray nozzle can help prevent soil erosion and damage to delicate plants.

Establishing a watering schedule tailored to your plant's needs and environmental conditions is critical. Watering in the morning is often recommended. Morning watering allows the plants to absorb moisture before the day's heat, reducing evaporation and helping them stay hydrated. Evening watering is an alternative, but avoiding wetting the foliage is essential to prevent fungal diseases. Adjusting the frequency of watering based on seasonal changes is also necessary. In summer, you may need to water daily or every other day, while in cooler months, once or twice a week may suffice. Monitoring soil moisture levels is a practical way to determine when your plants need water. Use a moisture meter or stick your finger into the soil until the second knuckles. If it feels dry at this depth, it's time to water.

Preventing watering issues in container gardening involves a few straightforward practices. Using mulch is an effective way to retain moisture. A layer of organic mulch, such as straw or wood chips, on the soil surface can reduce evaporation, keep the soil cool, and suppress weeds. Ensuring proper drainage is crucial to prevent root rot. Ensure your containers have adequate drainage holes and consider elevating them slightly off the ground to allow excess water to escape. Waterlogging can be avoided by not letting containers sit in saucers of water for extended periods. Elevating containers on pot feet or small bricks can help with this. These simple steps can prevent common watering problems and promote healthier, more vigorous plants.

Understanding the watering needs of different plants, employing efficient watering methods, and establishing a tailored watering schedule are all crucial aspects of successful container gardening. By paying attention to these details, you can create a thriving container garden that provides beauty and bounty year-round. Whether you use drip irrigation, self-watering containers, or hand watering, the goal is to ensure your plants receive the right amount of moisture to grow strong and healthy. Simple practices like mulching and ensuring proper drainage can significantly prevent common watering issues, allowing you to enjoy a vibrant and productive urban garden.

4.4 Vertical Gardening: Maximizing Space with Containers

Vertical gardening is an intelligent solution for urban gardeners looking to maximize their limited space. One of the most significant advantages of vertical gardening is its space efficiency. Using vertical space allows you to grow more plants in the same footprint, especially in small apartments or crowded urban areas. This method will enable you to transform walls, fences, and other vertical surfaces into lush, productive gardens. Vertical gardens add aesthetic appeal to your space. They can create stunning green walls and hanging gardens that not only beautify your surroundings but also provide a sense of tranquility. Improved air circulation and light exposure are additional benefits. Elevating plants off the ground allows for better airflow, reducing the risk of fungal diseases. Plants higher up also receive more light, which is beneficial in densely built urban areas where buildings can cast long shadows.

Various types of vertical gardening systems cater to different needs and preferences. Wall-mounted planters are ideal for herbs and small plants. These planters can be attached to walls or fences, creating a living tapestry of greenery. They are perfect for growing herbs like basil, thyme, and mint, which you can harvest regularly for cooking. Pocket gardens are another innovative solution. These fabric pockets are lightweight and can be easily mounted on walls. They are suitable for a variety of plants, including flowers like petunias, nasturtiums, and marigolds. Trellises and lattices, on the other hand, support climbing plants. Vegetables such as tomatoes, beans, and cucumbers thrive on these structures, making them an excellent choice for vertical gardening. The trellises support the plants in climbing, maximizing vertical space while keeping the plants off the ground and away from pests.

An image of planters attached to a fence with lush vegetables

When choosing plants for vertical gardens, consider those naturally thriving in upright positions. Herbs like basil, thyme, and mint are excellent choices. They grow well in shallow containers and can be harvested frequently. Flowers like petunias, nasturtiums, and marigolds add color and vibrancy to your vertical garden. These plants are easy to grow and maintain, making them ideal for beginners. Vegetables such as tomatoes, beans, and cucumbers are also well-suited for vertical gardening. These climbing plants use vertical space, producing abundant yields without taking up valuable ground area. When selecting plants, consider their growth habits and requirements to ensure they thrive in your vertical setup.

Installing and maintaining a vertical garden requires some planning and effort, but the results are well worth it. Securing planters and ensuring stability is the first step. Use sturdy brackets or hooks to attach wall-mounted planters and pocket gardens. If you use trellises, ensure they are firmly anchored to the ground or wall. Regular watering and fertilization are crucial for the health of your vertical garden. Because vertical gardens can dry out faster than traditional gardens, consider installing a drip irrigation system to provide consistent moisture. Alternatively, self-watering planters can help reduce the frequency of watering. Pruning and training climbing plants are essential maintenance tasks. Regularly trim back overgrown branches to maintain the shape and health of your plants. Train climbing plants to follow the trellis or lattice, ensuring they grow in the desired direction and do not become tangled.

You will need a few basic supplies to get started with vertical gardening. The essentials include:

- Wall-mounted planters or pocket gardens.

- A sturdy trellis or lattice.

- A selection of suitable plants.

Consider investing in a drip irrigation system or self-watering planters to simplify watering. Regularly monitor your vertical garden for signs of pests or diseases and take action promptly to address any issues. With some effort and creativity, you can transform any vertical space into a thriving, beautiful garden that enhances your living environment and connection to nature.

4.5 Best Vegetables and Herbs for Container Gardening

Selecting the best vegetables and herbs for container gardening ensures your urban garden thrives. Some vegetables are particularly well-suited for container gardening due to their space and growth requirements. Tomatoes are a favorite among urban gardeners. Cherry, dwarf, and patio varieties are perfect for containers. These smaller varieties produce abundant fruit and require less space than their larger counterparts. Leafy greens like lettuce, spinach, and kale are also excellent choices. They grow quickly, can be harvested multiple times, and do not need deep containers. Root vegetables, such as carrots, radishes, and beets, can also be grown in containers. They require deeper pots to accommodate their roots but are otherwise low-maintenance and highly productive.

Herbs are a fantastic addition to any container garden, providing fresh flavors for your culinary creations. Basil is an all-time favorite. It is easy to grow, versatile in the kitchen, and thrives in containers. Parsley is another excellent choice. It is compact, flavorful, and can be harvested continuously throughout the growing season. Rosemary stands out for its aromatic qualities and drought tolerance. This hardy herb is perfect for containers and adds a beautiful scent to your garden and dishes. Each of these herbs enhances your garden's productivity and brings a touch of greenery and freshness to your culinary endeavors.

Planting and caring for these vegetables and herbs require some attention to detail. Start by choosing the right container size and soil mix. Select a large, robust container with a depth of at least 18 inches for tomatoes. Use a rich soil mix with plenty of compost to provide nutrients. With a well-draining soil mix, leafy greens and herbs can be planted in shallower containers, around 6-12 inches deep. Root vegetables need deeper containers, at least 12 inches, to allow their roots to grow unimpeded. Regular fertilization is critical to maintaining healthy plants. Use a balanced, water-soluble fertilizer every two weeks during the growing season. Organic options like fish emulsion or seaweed extract are excellent choices. Pest management is also crucial. Keep an eye out for common pests like aphids and caterpillars. Use organic pest control methods like neem oil or insecticidal soap to keep them at bay without harming your plants or the environment.

Harvesting your vegetables and herbs immediately ensures optimal flavor and nutrition. For tomatoes, wait until the fruit is fully colored and slightly soft. Harvesting in the morning, when the fruit is excellent, helps retain its flavor. You can harvest leafy greens when the leaves are large enough to eat. Use scissors to snip the outer leaves, allowing the inner leaves to continue growing. Root vegetables should be harvested when they reach their mature size. Gently pull them from the soil, taking care not to damage the roots. Herbs are best harvested in the morning, just after the dew has dried. Their essential oils are most concentrated during this period, providing the best flavor. Snip the stems just above a leaf node to encourage fresh growth.

Once harvested, knowing how to store and use your produce can make all the difference. You can store tomatoes at room temperature until fully ripe, then refrigerate them to extend their shelf life. Leafy greens should be washed, dried, and stored in a perforated plastic bag in the refrigerator to keep them fresh. You can store root vegetables in a cool, dark place for several weeks. Herbs can be used fresh or dried for later use. To dry herbs, hang them in small bunches in a warm, airy place out of direct sunlight. Once dried, store them in airtight containers away from light and heat. Fresh herbs can also be frozen. Chop them and place them in ice cube trays with a bit of water or oil, then freeze. Use these herb cubes to add fresh flavor to soups, stews, and sauces.

Incorporating these vegetables and herbs into your cooking can elevate your meals. Fresh tomatoes can be sliced for salads, made into sauces, or roasted with olive oil and spices. Leafy greens are versatile, perfect for salads, stir-fries, or as a nutritious side dish. Root vegetables can be roasted, added to stews, or enjoyed raw in salads. Herbs like basil, parsley, and rosemary add fresh, vibrant flavors to various dishes. Try making a simple basil pesto, adding parsley to your salads, or using rosemary to flavor roasted vegetables and meats. The possibilities are endless, and the satisfaction of cooking with homegrown produce is unparalleled.

You can create a productive and beautiful urban garden by carefully selecting and caring for the best vegetables and herbs for container gardening. Each plant offers unique benefits that can be easily incorporated into daily life. Whether you are a seasoned gardener or just starting, these tips and techniques will help you grow a thriving container garden filled with fresh, delicious produce.

As we wrap up this chapter, remember that container gardening offers immense opportunities for creativity and productivity, even in limited spaces. The next chapter will delve into raised bed gardening techniques, providing more insights into optimizing your urban garden.

Chapter Four

Raised Bed Gardening Techniques

W hen I first built raised beds in my urban garden, I was over-whelmed by the choices and possibilities. The idea of creating a structured, elevated garden space was exciting, but I quickly realized that the materials and design choices I made would have a lasting impact on the success of my garden. Raised beds offer many benefits, including better control over soil quality, improved drainage, and easier access to planting and maintenance. This chapter explores how to design and build raised beds, focusing on choosing suitable materials, essential design considerations, step-by-step building techniques, and selecting the best location for your raised beds.

5.1 Designing and Building Raised Beds: Materials and Methods

Choosing suitable materials for constructing raised beds is crucial for durability and sustainability. Wood is among the most popular choices, particularly untreated cedar and redwood. These types of wood are naturally resistant to decay and insects, making them ideal for outdoor use. Reclaimed wood is another excellent option, offering an eco-friendly way to reuse materials while adding a rustic charm to your garden. Metal raised beds, made from galvanized steel or corrugated metal, are also gaining popularity. They are durable, long-lasting, and add a modern aesthetic to your garden. Stone options like concrete blocks and bricks provide a sturdy and permanent solution. These materials are excellent for creating a classic garden look and withstand the elements for many years. Compost materials combine recycled plastic and wood fibers and offer the best of both worlds. They resist decay, require minimal maintenance, and are environmentally friendly.

When designing your raised beds, several factors should be considered to ensure they meet your gardening needs. The size and shape of your raised beds will depend on the space and the plants you intend to grow. Standard dimensions for ease of access are typically 4 feet wide, allowing you to reach the center of the bed from either side without stepping into it. The length can vary based on your space, but typical lengths are 4, 8, or 12 feet. Modular designs offer flexibility, allowing you to expand or reconfigure your garden as needed. Elevated beds are handy for gardeners with mobility issues, as they reduce the need for bending and kneeling. These beds can be built higher off the ground, making them accessible for wheelchair users or those who find it challenging to work at ground level.

Building raised beds involves several steps, from planning to assembly. Start by measuring and cutting your chosen materials to the desired dimensions. If using wood, use untreated or non-toxic treated wood to avoid chemical leaching into the soil. Assemble the frames using screws or brackets for stability. Ensure structural stability by reinforcing the corners with corner supports. These can be made from wood, metal brackets, or even additional pieces of the same material used for the bed. Once the frame is assembled, place it in the desired location and level the ground to ensure even soil distribution. Fill the bed with a high-quality soil mix, as discussed in previous chapters, to provide a fertile growing environment for your plants.

Placement and positioning of your raised beds are critical for maximizing sunlight exposure and ensuring proper drainage. Most vegetables and flowers require at least 6-8 hours of direct sunlight daily, so choose a location with ample sunlight. Avoid placing your beds near structures or trees that could cast shadows and limit light exposure. It is also essential to consider the root systems of nearby trees, as they can compete with your garden plants for nutrients and water. Ensure proper drainage by leveling the ground before placing your raised bed. This practice prevents water from pooling at one end of the bed and promotes even moisture distribution throughout the soil. If your garden area has poor drainage, consider adding a layer of gravel or coarse sand beneath the raised bed to improve water flow.

Raised beds offer a versatile and productive gardening solution, especially for urban environments where space and soil quality may be limited. By carefully selecting materials, considering essential design aspects, following proper building techniques, and choosing the optimal location, you can create a thriving garden space that enhances your home and connection to nature.

Raised Bed Building Checklist

- Materials:

- ○ Wood: untreated cedar, redwood, and reclaimed wood

- ○ Metal: galvanized steel and corrugated metal

- ○ Stone: concrete blocks and bricks

- ○ Composite materials: recycled plastic and wood composites

- Design Considerations:

  - ○ Standard dimensions for ease of access (typically 4 feet wide)

  - ○ Modular designs for flexibility

  - ○ Elevated beds for gardeners with mobility issues

- Building Techniques:

  - ○ Measuring and cutting materials

  - ○ Assembling frames using screws or brackets

  - ○ Ensuring structural stability with corner support

- Placement and Positioning:

  - ○ Sunlight requirements for different plants

  - ○ Avoiding shaded areas and tree roots

  - ○ Ensuring proper drainage by leveling the ground

Choosing suitable materials and designs for your raised beds can make all the difference in the success of your urban garden. Whether you opt for wood, metal, stone, or composite materials, each offers unique benefits that can enhance your gardening experience. Following these guidelines, you can build durable, functional, and beautiful raised beds that will serve you well for years.

5.2 Soil Preparation for Raised Beds: Layering and Amending

A rich, fertile environment in your raised beds starts with adequate soil preparation. One of the best techniques for ensuring optimal soil health and structure is layering. This method mimics the natural layering found in forests and fields, providing a balanced and nutrient-rich environment for your plants. Begin with a base layer of coarse materials such as sticks, straws, or small branches. This layer promotes drainage, preventing water from pooling at the bottom of the bed and ensuring that plant roots receive adequate oxygen. On top of this, add an intermediate layer of organic matter like compost and leaf mold. This layer decomposes over time, enriching the soil with essential nutrients and improving its texture. Finally, top off your raised bed with high-quality garden soil and additional compost. This top layer should be the richest in nutrients, providing immediate nourishment for your plants.

Soil amendments are crucial in enhancing the fertility and structure of your raised bed soil. Adding compost is one of the most effective ways to increase organic matter content, improve soil structure, and boost nutrient levels. Compost provides a slow-release source of nutrients, feeding your plants over time and helping to retain moisture in the soil. Green manure and cover crops are also valuable soil amendments. These plants are grown specifically to be turned into the soil, adding organic matter and nutrients. You can plant cover crops like clover or rye in the off-season and till them into the soil before planting your main crops. Rock dust is another beneficial amendment. It supplies essential trace minerals that might be lacking in your soil, contributing to overall plant health and vitality.

Balancing soil pH is essential for optimal plant growth, as it affects nutrient availability. Testing your soil pH is straightforward. You can use a home testing kit or send a sample to a professional lab for detailed analysis. Most plants prefer a slightly acidic to neutral pH, typically between 6.0 and 7.0. If your soil is too acidic, adding lime can raise the pH. Lime should be applied in the fall or early spring, as it takes time to break down and affect the soil.

If your soil is too alkaline, adding sulfur can lower the pH. Sulfur is best applied in the fall, giving it time to work before the growing season begins. Regularly testing and adjusting your soil pH ensures plants can access the nutrients they need for healthy growth.

Improving soil structure is an ongoing process that requires regular attention. One of the fundamental practices is avoiding soil compaction. Compacted soil restricts root growth and reduces the availability of water and nutrients. To prevent this, avoid walking on your raised beds. Instead, use stepping stones or create designated paths to minimize soil disturbance. Regularly adding organic matter, such as compost or well-rotted manure, helps to maintain and improve soil structure. These materials break down over time, creating a loose, crumbly texture ideal for plant roots. Mulching is another effective technique for protecting soil structure. Apply a layer of mulch, such as straw, wood chips, or leaves, to the soil surface. Mulch helps retain moisture, suppress weeds, and gradually adds organic matter to the soil as it decomposes.

Following these soil preparation techniques can create a fertile, well-structured environment that supports healthy plant growth in your raised beds. Layering different materials ensures balanced drainage and nutrient availability, while regular amendments improve soil fertility and structure. Maintaining the proper soil pH and protecting soil structure through careful management is essential for a thriving garden. With these practices, your raised beds will become a productive and sustainable part of your urban gardening efforts, providing abundant harvests and vibrant plants.

5.3 Planting Strategies for Raised Beds: Maximizing Yield

Maximizing Yield in raised beds involves several strategic planting techniques that can make a significant difference. One such technique is companion planting, which pairs plants that benefit each other. This method not only maximizes space but also improves plant health and productivity. For example, tomatoes and basil are a classic pair. Basil repels pests that commonly afflict tomatoes, while tomatoes provide a shade that helps basil thrive in hotter climates. Another effective combination is carrots and onions. Carrots attract beneficial insects that prey on pests harmful to onions, and onions deter carrot flies. By carefully selecting and pairing plants, you can create a symbiotic environment that enhances growth and yield.

Succession planting is another powerful strategy to ensure a continuous harvest throughout the growing season. This technique involves planting fast-growing crops immediately after harvesting early-season crops, ensuring no space goes to waste. For instance, after harvesting early lettuce in the spring, you can grow bush beans that will mature by late summer. Staggering planting times are also crucial to extended yields. Planting crops like radishes or leafy greens in two-week intervals allows you to harvest continuously rather than all at once. This approach not only maximizes the use of your raised beds but also provides a steady supply of fresh produce.

Square-foot gardening is a method that optimizes space in raised beds by dividing the area into smaller sections, typically one square foot each. This intensive planting technique allows you to grow crops in a compact space, making it perfect for urban gardens. Each square foot can be planted with a different crop, depending on its spacing requirements. For example, one square foot can hold one tomato plant, four lettuce plants, or sixteen carrots. This method encourages dense planting, which helps suppress weeds and retain soil moisture. Using a grid system, you can plan and plant your raised bed efficiently, ensuring that every inch is utilized effectively.

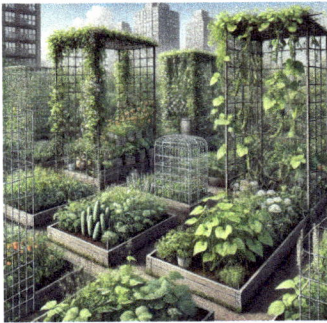

Image showing a lush urban garden with trellises, cages and climbing plants

Incorporating vertical growing techniques within raised beds is another excellent way to maximize space. Using trellises and cages for climbing plants like cucumbers, peas, and beans allows you to grow vertically, saving precious horizontal space for other crops. Training plants upward maximizes space, improves air circulation, and reduces disease risk. For instance, installing a sturdy trellis at the back of your raised bed and planting climbing beans or cucumbers can create a vertical garden that frees up space for low-growing plants like lettuce or carrots in the front. This multi-layer approach enhances productivity and makes the most of your available space.

Combining these strategies—companion planting, succession planting, square-foot gardening, and vertical growing—can significantly boost the Yield of your raised beds. By carefully planning and implementing these techniques, you create a dynamic and efficient garden space that maximizes productivity and sustainability. Each method complements the others, creating a harmonious environment where plants thrive and flourish. Whether growing vegetables, herbs, or flowers, these planting strategies offer an innovative, effective way to make the most of your raised beds.

Using these methods, you can transform your raised beds into highly productive and sustainable garden spaces. Companion planting fosters beneficial relationships between plants, while succession planting ensures continuous harvests. Square-foot gardening maximizes space through intensive planting, and vertical growing techniques make the most of limited horizontal space. Each strategy enhances the overall productivity of your garden, providing you with abundant, healthy produce throughout the growing season.

These planting techniques are not only practical but also rewarding. Watching your garden flourish because of intelligent planning and efficient use of space brings immense satisfaction. Combining these methods creates a self-sustaining ecosystem where plants support each other and thrive together. With careful attention to detail and thoughtful planning, your raised beds can become a model of productivity and sustainability, offering a continuous supply of fresh, nutritious produce.

5.4 Irrigation Solutions for Raised Beds

Efficient watering is crucial for the success of raised bed gardens, and one of the best methods to achieve this is through drip irrigation systems. Setting up a drip irrigation system involves installing drip lines and emitters to deliver water directly to the plant roots. Start by laying out the main water supply line, then connect drip lines to this main line. Place emitters at the base of each plant or evenly spaced along rows to ensure consistent watering. Adjusting watering schedules based on plant needs is essential. For example, young seedlings may require more frequent watering, while established plants need deeper, less frequent irrigation. The benefits of drip irrigation are numerous: it conserves water by minimizing evaporation and runoff and provides targeted watering, ensuring that each plant receives the right amount of moisture.

Soaker hoses offer another effective irrigation method for raised beds. These porous hoses release water slowly along their entire length, making them ideal for even water distribution. To install soaker hoses, lay them out along the rows of your raised beds, ensuring they are close to the plant roots. Covering the hoses with mulch helps retain moisture and reduce water evaporation. This method is easy to install and maintain, ensuring water penetrates deeply into the soil, promoting healthy root growth.

Self-watering systems benefit busy gardeners needing more time to water their plants regularly. Wicking beds are a popular self-watering system that uses capillary action to draw water from a reservoir into the soil. To set up a wicking bed, create a water reservoir at the bottom of your raised bed using a layer of stones or gravel, then place a layer of fabric over the reservoir to act as a wick. Fill the bed with soil, ensuring that the wick reaches up into the root zone of your plants. The water reservoir can be refilled as needed, providing a consistent source of moisture. These systems are highly efficient, reducing the watering frequency and ensuring that plants receive a steady water supply.

Manual watering remains a practical method for many gardeners. Best practices for manually watering raised beds include watering in the morning to reduce evaporation and using a watering can with a fine nozzle to avoid soil compaction and waterlogging. Monitoring soil moisture is crucial to prevent over-watering, which can lead to root rot. Checking the soil's moisture level by sticking your finger into the soil can help determine if watering is needed. Aim to keep the soil moist but not waterlogged, ensuring that plants receive the right amount of water for optimal growth.

These irrigation solutions can significantly improve the efficiency and effectiveness of watering your raised beds. Drip irrigation systems provide targeted watering, conserving water and ensuring that each plant receives consistent moisture. Soaker hoses offer even water distribution and are easy to install and maintain. Self-watering systems, such as wicking beds, provide a consistent source of moisture and reduce watering frequency, making them ideal for busy gardeners. When done correctly, manual watering ensures even moisture levels and allows for close monitoring of plant health.

Incorporating these irrigation methods into your raised bed gardening routine can lead to healthier plants and more productive gardens. By understanding the specific needs of your plants and choosing the right irrigation system, you can create an efficient and effective watering regimen that supports robust plant growth. Whether you choose drip irrigation, soaker hoses, self-watering systems, or manual watering, each method offers unique benefits that you can tailor to your gardening needs. Efficient watering conserves resources and promotes healthier, more productive plants, ensuring your raised beds thrive throughout the growing season.

5.5 Seasonal Maintenance and Crop Rotation in Raised Beds

Maintaining raised beds throughout the seasons involves specific tasks that ensure your garden remains productive year-round. It is time to prepare the soil and plant the early crops in spring. Begin by clearing any winter debris and checking the soil's moisture level. If the soil is too wet, wait for it to dry out to avoid compaction. Add a fresh layer of compost to replenish nutrients and improve soil structure. This period is also the perfect time to plant cool-season crops like lettuce, spinach, and peas. These plants thrive in the cooler temperatures of early spring and can be harvested before the summer heat sets in.

During the summer, focus on mulching and pest control to keep your garden thriving. Mulch helps retain soil moisture, suppresses weeds, and regulates soil temperature. Apply a thick layer of organic mulch, such as straw or wood chips, around your plants. This practice conserves water and reduces the need for frequent weeding. Summer is also prime time for pests, so regular monitoring is crucial. Look for early signs of pest activity, such as chewed leaves or discolored spots. Use organic pest control methods like neem oil or insecticidal soap to manage infestations without harming beneficial insects.

As fall approaches, adding compost and planting cover crops is essential to preparing your raised beds for the off-season. Turn in any remaining summer crops and add a generous layer of compost to enrich the soil. Plant cover crops like clover, rye, or buckwheat. When decomposed, these plants protect the soil from erosion, suppress weeds, and add organic matter. Cover crops can be tilled into the soil in the spring, providing a nutrient boost for your next planting cycle.

Protect your raised beds with mulch or covers in winter to preserve soil integrity. Apply a thick layer of mulch to insulate the soil and protect it from freezing temperatures. Alternatively, you can use row covers or garden fabric to shield your beds from harsh weather.

This practice helps maintain soil structure and prevents nutrient loss. Winter is also an excellent time to plan for the next growing season, reviewing your successes and challenges from the past year and adjusting as needed.

Crop rotation is vital for maintaining soil health and preventing pest buildup in raised beds. Rotating families of crops—such as legumes, brassicas, and root vegetables—helps disrupt the life cycles of pests and diseases. For example, legumes like beans and peas fix nitrogen in the soil, benefiting subsequent crops like leafy greens that require high nitrogen levels. Brassicas, such as broccoli and cabbage, should follow legumes, as they thrive on the residual nitrogen. You can plant root vegetables like carrots and beets after brassicas to take advantage of the remaining nutrients. By rotating these crop families, you reduce the risk of soil depletion and minimize pest and disease problems.

Cover cropping is another effective method for maintaining soil fertility and structure during off-seasons. Choose suitable cover crops like clover, rye, or buckwheat, which are known for their soil-enhancing properties. These crops proliferate, providing ground cover that prevents erosion and suppresses weeds. Once they mature, cut them down and turn them under to incorporate them into the soil. This process adds organic matter and improves soil structure, making it fertile for future plantings.

Managing pests and diseases throughout the seasons requires vigilance and proactive strategies. Regularly monitor your garden for early signs of pests. Look for common indicators like holes in leaves, discolored spots, or stunted growth. Early detection allows you to address problems before they escalate. Implement organic pest control methods, such as introducing beneficial insects like ladybugs or using natural repellents like garlic spray. Crop rotation and intercropping are also effective in disrupting pest cycles. Intercropping involves planting different crops together to confuse pests and reduce their impact. For instance, planting marigolds alongside tomatoes can deter nematodes and other harmful insects.

You can keep your raised beds productive and healthy year-round by diligently performing seasonal maintenance tasks, practicing crop rotation, and employing cover cropping. Each season brings activities that contribute to the overall success of your garden. From preparing the soil in spring to protecting it in winter, these practices ensure that your raised beds remain fertile and vibrant. In the next chapter, you will learn more about integrating these techniques with advanced soil management practices to enhance your urban garden's productivity and sustainability.

Chapter Five

Soil Preparation and Management

I was eager to plant and watch seeds grow when I began urban gardening. However, I quickly realized that my enthusiasm alone was not enough. My plants were struggling, and I could not understand why. Things changed when I learned about the importance of soil testing. Soil testing became a significant change for me. It allowed me to understand what my soil needed and how to provide it. This chapter is about equipping you with the knowledge and tools to prepare and manage your soil effectively, ensuring your urban garden thrives.

3.1 Soil Testing: Techniques and Tools for Urban Gardeners

Regular soil testing is crucial for urban gardens. It is like a health check-up for your soil, helping you identify what it needs to support healthy plant growth. One of the primary reasons for soil testing is to identify nutrient deficiencies. For plants to flourish, they require a range of nutrients; any deficiency can cause stunted growth and lower yields. For example, nitrogen deficiency can cause yellowing leaves, while a lack of phosphorus may cause stunted growth. Soil testing accurately identifies nutrient shortfalls, enabling targeted correction through fertilizers or soil enhancements.

Detecting pH imbalances is another critical aspect of soil testing. Soil pH affects nutrient availability and can significantly impact plant health. Most plants prefer a slightly acidic to neutral pH, typically between 6.0 and 7.0. If your soil is too acidic or alkaline, certain nutrients become less available to plants, leading to deficiencies even if those nutrients are present in the soil. For instance, in highly acidic soil, essential nutrients like phosphorus and magnesium become less accessible. You can make necessary adjustments to create an optimal growing environment by testing your soil's pH.

Soil testing also uncovers potential contaminants, which is essential in urban settings. Pollutants like heavy metals and chemicals can pose severe risks to plants and humans. Lead, for instance, is a common contaminant in urban soil because of past use in paint and gasoline. Awareness of the levels of such pollutants empowers you to take the proper remediation measures, guaranteeing a safe environment for growing edible plants in your garden.

There are several types of soil tests available to gardeners. pH testing kits are among the most common and easiest to use. These kits usually include a pH meter or test strips that change color based on the soil's acidity or alkalinity. Nutrient test kits are another valuable tool, providing insights into the levels of critical nutrients like nitrogen, phosphorus, and potassium. These kits often involve mixing soil samples with a solution and comparing the color change to a chart. For those concerned about contaminants, contaminant screening kits are available. These kits can test for heavy metals and other harmful substances, providing peace of mind that your soil is safe.

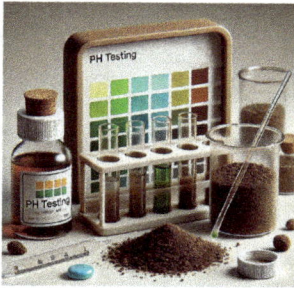

Image showing a pH testing kit for soil analysis

Using soil testing kits effectively involves a few essential steps. Start by collecting soil samples from different parts of your garden. This practice ensures a representative sample, as soil conditions vary within a small area. Use a clean trowel to dig small holes about six inches deep and take samples from the sides of the holes. Mix these samples in a clean bucket and remove debris like stones or roots. Follow the instructions on your test kit carefully. PH testing usually involves mixing soil with distilled water and using a pH meter or test strip to measure the alkalinity or acidity. Nutrient and contaminant tests may require mixing soil with a reagent solution and waiting for a color change. Interpreting the results involves comparing the test outcomes to standard charts provided with the kits, which will guide you on necessary amendments.

While home testing kits are convenient, professional soil testing services offer a more comprehensive analysis. Local agricultural extension services often provide soil testing at a reasonable cost. These services can test for a broader range of nutrients and contaminants, offering detailed insights into your soil's health. Certified soil laboratories are another option, providing in-depth analysis and specific recommendations for soil amendments. These professional tests are beneficial if you suspect significant contamination or need precise nutrient management.

Soil testing is fundamental in urban gardening, providing the information needed to create a healthy growing environment. Regularly testing your soil can identify nutrient deficiencies, detect pH imbalances, and uncover potential contaminants. This knowledge lets you make informed decisions about soil amendments and create a thriving urban garden. Whether you use home testing kits or professional services, soil testing is an investment in the success of your garden.

3.2 Amending Soil: Organic Matter and Soil Conditioners

Organic matter is one of the most transformative elements you can introduce to your urban garden. It is the backbone of healthy soil, improving its structure, fertility, and overall vitality. Organic matter plays a crucial role in enhancing soil aeration. When you add organic materials, they break down and create spaces in the soil, allowing air to reach plant roots and beneficial microorganisms. This increased aeration is vital for root health and overall plant vigor.

Organic matter improves water retention, acting like a sponge that holds moisture and gradually releases it to the plants. This is beneficial in urban environments, where soil can dry out quickly due to limited space and exposure to heat. Organic matter also boosts microbial activity in the soil. Microorganisms break down organic material into nutrients plants can easily absorb, creating a dynamic and fertile growing environment.

Several types of organic matter are suitable for urban gardens, each offering unique benefits. Compost is the most versatile and widely used. It is made from decomposed kitchen scraps, garden waste, and other organic materials. The manure enriches the soil with essential nutrients and improves its structure. Leaf mold, created from decomposed leaves, is another excellent option. It is rich in humus, which enhances soil fertility and water retention. Aged manure, sourced from farm animals like cows and horses, is also highly beneficial. It provides a slow-release source of nutrients and improves soil texture. You can use these types of organic matter alone or combine them to create a fertile soil mix.

Besides organic matter, soil conditioners are valuable tools for improving soil quality in urban gardens. Gypsum effectively breaks up clay soils, making them more workable and improving drainage. It is beneficial in areas with heavy, compacted soil where water pools. Perlite and vermiculite are lightweight volcanic minerals that enhance soil aeration and drainage. They are especially helpful in container gardening, where soil compaction can be a significant issue. Coconut coir, made from the fibers of coconut husks, is an excellent alternative to peat moss. It enhances water retention and provides a stable, airy structure for plant roots. Using these soil conditioners in combination with organic matter can create an optimal growing environment for your plants.

Correctly applying organic matter and soil conditioners is critical to maximizing their benefits. Start by mixing organic matter into the topsoil. A general rule of thumb is to add a 2–3-inch layer of compost or other organic materials to the surface and then work it into the top 6-8 inches of soil. This practice ensures that organic matter is well-incorporated and readily available to plant roots. Creating compost tea is another effective method. Compost tea is a liquid extract made by steeping compost in water, which you can use to water your plants. It provides a quick boost of nutrients and beneficial microorganisms. To make compost tea, fill a container with water, add a bag of compost, and let it steep for 24-48 hours. Strain the liquid and use it to water your plants.

Layering techniques are beneficial for raised beds. To improve drainage, add a base layer of coarse materials like straw or small branches. Next, add alternating layers of green materials (kitchen scraps, fresh grass clippings) and brown materials (dried leaves, cardboard) to create a rich, compost-like mix. Top it off with a layer of high-quality soil or compost. This technique enriches the soil, helps retain moisture, and suppresses weeds.

Incorporating these practices into your urban garden can improve soil health and plant growth. Organic matter and soil conditioners work together to create a fertile, well-structured soil that supports robust plant development. Whether growing vegetables in containers, flowers in raised beds, or herbs on a windowsill, understanding and applying these principles will help you cultivate a thriving urban garden.

3.3 Composting in Small Spaces: Methods and Benefits

Composting in urban environments offers many benefits, turning kitchen waste into a valuable resource for your garden. One of the most immediate advantages is reducing kitchen waste. You cut down on landfill waste when composting vegetable scraps, coffee grounds, or eggshells. This practice helps the environment and reduces methane emissions, a potent greenhouse gas. Composting produces nutrient-rich compost that can enrich your garden soil. This nutrient boost enhances soil health, making it more fertile and capable of supporting robust plant growth. Healthy soil leads to healthier plants more resistant to pests and diseases.

Regarding composting in small urban spaces, several practical and space-efficient methods exist. Vermicomposting with worm bins is a popular choice. This method uses red wigglers or other composting worms to break down organic matter into nutrient-rich worm castings. Worm bins are compact and can be kept indoors or on a balcony, making them ideal for apartment dwellers. Another option is Bokashi composting, a fermentation process that uses beneficial microbes to break down kitchen scraps. Bokashi bins are airtight and can be kept under the kitchen sink without odor issues. The process is quick, typically taking about two weeks, and the resulting material can be buried in the garden or added to a compost pile. Tumbler composters are also an excellent choice for small spaces. These enclosed bins can be rotated to aerate the compost, accelerating the decomposition process. Tumbler composters are efficient, reducing composting time to just a few weeks, and they are easy to use, requiring minimal effort to maintain.

Effective composting requires thoroughly understanding what materials can and cannot be composted. Green materials, including vegetable scraps, coffee grounds, and fresh grass clippings, are nitrogen-rich. These materials provide the nitrogen necessary for microbial activity and help speed up composting. Brown materials, rich in carbon, include dried leaves, cardboard, and shredded paper. Carbon is essential for providing energy to microorganisms, balancing the high nitrogen content of green materials. Avoid composting meat, dairy, and oily foods, as these can create odors and attract pests. Refrain from adding diseased plants or weeds that have gone to seed, as they can introduce pathogens and unwanted seeds into your compost.

Maintaining a compost bin in a small space requires attention to detail to ensure efficient decomposition. Aeration is key. Compost needs oxygen to break down organic matter effectively. Regularly turning the compost or rotating a tumbler composter helps introduce air, preventing the compost from becoming anaerobic and smelly.

Balancing green and brown materials is another critical factor. Aim for a ratio of about one part green to three parts brown. This balance ensures that compost has enough nitrogen for microbial activity and enough carbon to provide energy. Monitoring moisture levels is also crucial. The compost should be moist, like a wrung-out sponge, but not waterlogged. Too much moisture can lead to anaerobic conditions, while too little can slow decomposition. If your compost becomes too dry, add more green materials or water. If it's too wet, add more brown materials to absorb the excess moisture.

You can successfully compost in even the smallest urban spaces by understanding these principles and methods. Composting reduces waste, produces valuable soil amendments, and fosters a deeper connection with your garden and the natural cycles of growth and decay. Whether you choose vermicomposting, Bokashi, or a tumbler composter, each method offers unique advantages that can be tailored to your specific needs and living situation. With attention and care, your compost bin will become an indispensable part of your urban gardening routine, contributing to healthier soil and more vibrant plants.

3.4 Vermiculture: Using Worms to Enhance Soil Quality

Vermiculture, or worm composting, is a significant change for urban gardeners. It involves using worms to decompose organic waste, producing high-quality worm castings that enrich the soil. Worm castings, also known as vermicompost, are rich in nutrients and beneficial microorganisms, making them an excellent soil amendment. This process accelerates composting and reduces kitchen waste, making it a valuable resource for your garden.

Setting up a worm bin is straightforward and can be done in small spaces:

1. Choose the correct type of bin. You can purchase a commercial worm bin or make one using a plastic container. Ensure the bin has a tight-fitting lid and drill small holes for ventilation and drainage.

2. Prepare the bedding material. Shredded newspaper, cardboard, and coconut coir are excellent options. Moisten the bedding to make it feel like a damp sponge, then add a handful of garden soil to introduce beneficial microorganisms.

3. Add the worms.

Red wigglers are the best for vermiculture because they efficiently break down organic matter. You can purchase them online or from a local supplier. Add the worms to the bin and let them settle into their new home.

Feeding your worms is simple, but it requires some care. Suitable food scraps include vegetable peels, fruit scraps, coffee grounds, and eggshells. These provide the necessary nutrients for the worms to thrive. However, avoid feeding them citrus, onions, and spicy foods, as these can create an acidic environment that harms the worms. Also, steer clear of meat, dairy, and oily foods, which can attract pests and produce odors. Feed the worms in small amounts, gradually increasing the quantity as they adjust to their environment. Bury the food scraps under the bedding to minimize odors and discourage fruit flies.

Harvesting worm castings is a rewarding part of vermiculture. After a
few months, you will notice a rich,
dark compost accumulating in the
bin. To harvest the castings, you can
use the "light method." Move the bedding and castings to one side of the
bin, add fresh bedding and food to the
empty side, and expose the bin to light.
Worms dislike light and will migrate to

*An image showing vermicom-
posting in a bin for the urban
garden*

the new bedding, allowing you to collect the castings from the other
side. Another method is to use a mesh screen to separate the worms
from the castings. Once harvested, the worm castings can be mixed
into your garden soil, providing a nutrient boost. You can also use it
to create worm-casting tea. To make the tea, place a handful of castings
in a cloth bag, steep it in water for 24 hours, and use the liquid to
water your plants. This tea provides an immediate nutrient boost and
enhances microbial activity in the soil.

Worm castings significantly improve soil structure and fertility.
They enhance soil aeration and water retention, making the soil more
hospitable for plant roots. The beneficial microorganisms in the castings help break down organic matter, releasing nutrients that plants
can readily absorb. This process creates a dynamic, healthy environment supporting robust plant growth.

Vermiculture is a sustainable and efficient way to manage kitchen
waste and improve soil quality in urban gardens. It requires minimal
space and effort but yields substantial benefits. By setting up a worm
bin, feeding the worms appropriately, and harvesting the castings, you
can create a continuous supply of nutrient-rich compost for your

garden. Whether growing vegetables in containers, herbs on a windowsill, or flowers in raised beds, vermiculture can enhance your soil and promote healthy plants.

3.5 Creating the Perfect Soil Mix for Containers and Raised Beds

Creating the perfect soil mix for your urban garden is fundamental to the health and productivity of your plants. A good soil mix comprises several vital components that provide nutrients, retain water, and ensure proper drainage. Compost is an essential ingredient supplying the nutrients for plant growth. It is rich in organic matter, which helps improve soil structure and feeds beneficial microorganisms. Peat moss or coconut coir is crucial for water retention. These materials hold moisture well, ensuring that your plants have consistent access to water, even in containers where soil tends to dry out quickly. Finally, perlite or sand is added to the mix to enhance drainage. These lightweight materials prevent soil compaction and allow excess water to drain away, reducing the risk of root rot.

Fresh soil mixes cater to specific gardening needs. A general-purpose blend is versatile and suitable for most plants. This mix typically comprises equal parts compost, peat moss (or coconut coir), and perlite. The compost provides nutrients, the peat moss retains moisture, and the perlite ensures good drainage. For starting seeds, a finer-textured mix with more perlite is ideal. This seed starting mix allows delicate roots to penetrate quickly and ensures they are not waterlogged. Heavy feeder plants, like tomatoes and peppers, benefit from a richer mix with additional compost and organic fertilizers. This heavy feeder mix supplies these plants with extra nutrients for vigorous growth and high yields.

Customizing soil mixes based on the specific needs of your plants can make a significant difference in their health and productivity. Acid-loving plants, such as blueberries and azaleas, thrive in more acidic soil. To create a suitable mix for these plants, you can add sulfur or pine needles to lower the pH. Succulents and cacti require well-draining soil to prevent root rot. For these plants, increase the proportion of sand in your mix and reduce the amount of organic matter. This practice creates a gritty, fast-draining soil that mimics their natural habitat. Understanding the specific needs of your plants allows you to tailor your soil mix, providing the ideal conditions for each variety.

Maintaining the quality of your soil mix over time is crucial for ongoing plant health. Regularly adding compost helps replenish nutrients and organic matter, keeping the soil fertile and well-structured. Avoid over-fertilization, as it can lead to nutrient imbalances and harm your plants. Instead, follow recommended fertilizing guidelines based on your plant's specific needs and the results of your soil tests. Replacing the soil every few years is also essential, especially in containers. Over time, soil can become compacted and depleted of nutrients. Refreshing the soil ensures that your plants have access to the nutrients and structure they need to thrive.

Incorporating these practices into your urban gardening routine can significantly enhance your plants' health and productivity. You can create an optimal growing environment by understanding the components of a good soil mix and customizing it for specific plants. Maintaining soil quality through regular amendments and careful management further ensures that your garden remains productive year after year. Whether you grow vegetables in containers or flowers in raised beds, the right soil mix is critical to your success.

Creating the perfect soil mix involves understanding your plants'
needs and the characteristics of different components. Compost, peat
moss, and perlite work together to provide nutrients, retain moisture,
and ensure good drainage. Customizing your mix for specific plants,
such as acid-loving varieties or succulents, allows you to cater to their
unique requirements. Regular maintenance, including composting
and replacing soil, keeps your mix fertile and well-structured. With
these principles, you can create a thriving urban garden, no matter the
space constraints.

Chapter Six

Organic Pest and Disease Management

The sun was beginning to set as I stepped into my small urban garden, only to find my beloved basil plants covered in tiny, green insects. My heart sank. Like many urban gardeners, the ongoing battle with pests frustrated me. Determined to find an eco-friendly solution, I dove into organic pest management. This chapter is dedicated to sharing what I learned, helping you protect your garden without resorting to harmful chemicals.

6.1 Identifying Common Urban Garden Pests

One of the first steps in organically managing pests is knowing precisely what you are up against in your garden. Urban gardens, regardless of their location, often face a handful of common pests. Aphids are one of the most frequent culprits. These small, soft-bodied insects come in various colors, including green, black, and yellow. They tend to congregate on the undersides of leaves and new growth, causing leaves to curl and wilt. Aphids also produce a sticky residue known as honeydew, which can attract ants and lead to a sooty mold.

Whiteflies are another common pest. These tiny, white-winged insects resemble miniature moths and are often found on the undersides of leaves. When disturbed, they fly up in a cloud, making them easy to identify. Whiteflies can cause stunted growth and yellowing leaves, weakening the plant.

Slugs and snails are notorious for their nocturnal munching habits. These mollusks leave large, irregular holes in leaves and can devastate young seedlings overnight. They thrive in damp, shady areas, making them familiar in gardens with dense foliage or moisture-retentive soil.

Spider mites are tiny arachnids that can be challenging to see with the naked eye. They gather on the undersides of leaves, forming colonies. A telltale sign of spider mites is the fine silk webbing they produce. If left unchecked, they can cause leaves to turn yellow and cause significant damage.

Caterpillars, the larvae of moths and butterflies, are also frequent garden visitors. While beautiful as butterflies, their larvae can be quite destructive. Caterpillars chew large holes in leaves and quickly strip a plant of its foliage. They come in various sizes and colors, often blending with the leaves they feed on.

Recognizing the early signs of pest infestation can save your plants from severe damage. Yellowing or distorted leaves often indicate the presence of pests like aphids or spider mites. Sticky residue on leaves, known as honeydew, is a clear sign of aphids or whiteflies. Caterpillars, slugs, and snails typically cause holes in leaves and fruits. Regularly inspecting your plants for these signs can help you catch infestations before they become severe.

Monitoring your garden for pest activity is crucial for early detection and control. One effective technique is inspecting the undersides of leaves, where many pests like to hide and feed. This simple habit can help you spot aphids, whiteflies, and spider mites early on. Using yellow sticky traps is another helpful method. These traps attract and capture flying insects like whiteflies, allowing you to monitor their presence and act if necessary. Early morning garden walks can be particularly revealing for slug and snail activity. These pests are most active at night and early morning, leaving a slime trail in their wake. By checking your garden during these times, you can spot and remove them before they cause significant damage.

Pest Monitoring Checklist

- Inspect the Undersides of Leaves: Look for tiny insects, eggs, and webbing.

- Use Yellow Sticky Traps: Place traps near susceptible plants to catch flying insects.

- Early Morning Walks: Check for slugs and snails before they retreat to hiding spots.

- Look for Signs of Infestation: Yellowing leaves, sticky residue, and holes in foliage.

By understanding the common pests that plague urban gardens and regularly monitoring your plants, you can take proactive steps to protect your garden. Recognizing the early signs of infestation and knowing where to look can make all the difference in maintaining a healthy, thriving garden.

6.2 Natural Predators: Beneficial Insects for Pest Control

It was a revelation when I first discovered the concept of using beneficial insects for pest control. Instead of battling pests with chemicals, I could enlist the help of nature's defenders. Beneficial insects are a sustainable and effective way to manage pests in your garden. Ladybugs are a prime example. These vibrant beetles are voracious predators of aphids. A single ladybug can consume up to 50 aphids daily, making them an invaluable ally in the garden. They also eat other soft-bodied pests like mites and scales, contributing to a healthier garden ecosystem.

Lacewings are another beneficial insect worth welcoming into your garden. Often referred to as "aphid lions" in their larval stage, lacewings feed on various pests, including aphids, caterpillars, and whiteflies. They are particularly effective because they continue to prey on pests throughout their life cycle. The adults feed on nectar and pollen, making them excellent pollinators. Although tiny and often unnoticed, parasitic wasps are crucial in pest control. These wasps lay their eggs inside caterpillars and whiteflies. When the eggs hatch, the larvae feed on the host, effectively controlling the pest population. Despite their small size, parasitic wasps can significantly reduce pest numbers, making them a powerful tool in your pest management arsenal.

Attracting and maintaining populations of these beneficial insects requires creating a welcoming environment. Planting nectar-rich flowers like marigolds and Alyssum can attract ladybugs and lacewings to your garden. These flowers provide a food source for the adult insects, encouraging them to stay and reproduce. Marigolds, with their bright colors and strong scent, are particularly effective at attracting beneficial insects. With its sweet-smelling blooms, Alyssum attracts lacewings and draws in hoverflies, another beneficial insect that preys on aphids.

Providing a habitat is equally important. Insect hotels, structures made from natural wood and straw, offer shelter for beneficial insects. You can easily purchase or create a hotel at home, providing a safe place for insects to lay eggs and overwinter. Leaving some areas of your garden undisturbed also helps. Piles of leaves or straw can serve as nesting sites for ground beetles and other beneficial insects. Creating a diverse and welcoming environment encourages beneficial insects to reside in your garden, enhancing its natural defenses.

Sometimes, you may need to introduce beneficial insects to your garden to kick-start their population. Garden centers and online suppliers offer a variety of beneficial insects for sale. Ladybugs, lacewings, and parasitic wasps are commonly available and can be shipped directly to your home. When releasing these insects, timing and conditions are crucial. Release them in the early morning or late evening when temperatures are cooler, and they are less likely to fly away. Lightly misting the plants with water can also help, encouraging the insects to stay and explore their new environment. Distribute them evenly throughout your garden, focusing on areas with high pest activity.

Monitoring the effectiveness of beneficial insects in controlling pest populations is essential to ensure they are doing their job. Regularly inspect your plants for signs of pest activity. Look for reductions in pest numbers and improvements in plant health. Observing beneficial insect activity is also essential. Watch for ladybugs on your plants and lacewing larvae hunting for prey. Notice that pest populations are still high despite the presence of beneficial insects. Introducing more or adjusting your garden practices may be necessary to support them better. Keeping a garden journal can help track these observations and adapt as needed.

6.3 Homemade Organic Sprays and Remedies

Using homemade organic sprays and remedies offers significant advantages over chemical pesticides. These eco-friendly and non-toxic solutions ensure your garden remains safe for you, your family, and the environment. Unlike chemical pesticides, which can harm beneficial insects and contaminate soil and water, organic sprays target pests without leaving harmful residues. Moreover, they are cost-effective and easy to make, allowing you to use common household ingredients to protect your plants. This practice saves money and reduces reliance on store-bought products, promoting a more sustainable gardening practice.

One of the most versatile homemade sprays is a neem oil solution, effective for controlling various pests. Neem oil, derived from the neem tree, acts as an insect repellent, disrupts pests' life cycles, and inhibits feeding. To make a neem oil spray, mix two tablespoons of neem oil with a gallon of water and add a few drops of liquid soap to help the oil mix with the water. Use this spray on affected plants to deter aphids, whiteflies, and other common pests.

Garlic and chili spray are other powerful remedies for repelling insects. The strong scents of garlic and chili deter many pests, keeping your plants safe. To make this spray, blend a few cloves of garlic, two tablespoons of chili powder (or a handful of fresh chili peppers), and a quart of water. Let the mixture sit for a few hours, then strain it and add a few drops of liquid soap. This spray works well against aphids, spider mites, and caterpillars.

A simple soap and water solution can be highly effective for soft-bodied insects like aphids and whiteflies. The soap disrupts insects' cell membranes, causing them to dehydrate and die. Mix one tablespoon of liquid dish soap with a quart of water to make this solution. Spray it directly on the affected plants, ensuring thorough coverage of the leaves' tops and undersides.

Correctly applying these homemade remedies maximizes their effectiveness. The best times for application are early morning or late evening. This timing helps prevent the sun from burning the plants and allows the solution to work longer before evaporating. The frequency of application depends on the severity of the infestation. For mild infestations, spraying once a week is usually sufficient. For more severe issues, you may need to apply the solution every few days until the pest is controlled.

Safety is paramount when using homemade sprays. Always test the solution on a small area of the plant before a complete application to ensure it does not cause damage. Proper storage and labeling of homemade solutions are also essential. Please keep them in a cool, dark place, and label each container with its contents and the date you made it. This practice helps ensure you use the solution within its effective period and avoid mix-ups. Avoid spraying during high temperatures, as this can cause the solution to evaporate quickly and potentially burn the plants. If you use garlic and chili spray, wear gloves and avoid touching your face to prevent irritation.

The benefits of using homemade organic sprays extend beyond pest control. These remedies contribute to a healthier garden ecosystem by preserving beneficial insects and microorganisms. They also encourage a more hands-on approach to gardening, fostering a deeper connection with your plants and their needs. Furthermore, using natural ingredients reduces your carbon footprint and promotes sustainability in your gardening practices.

6.4 Disease Prevention: Soil Health and Plant Care

Healthy soil is the foundation of a disease-free garden. Maintaining soil health can prevent many plant diseases by promoting beneficial microorganisms and enhancing nutrient uptake. Beneficial microorganisms, such as bacteria and fungi, break down organic matter, release nutrients, and protect plants from pathogens. These microorganisms create a dynamic ecosystem in the soil that supports robust plant growth and resilience against diseases. When the soil is healthy, plants can absorb nutrients more efficiently, making them stronger and more disease-resistant.

Regularly add compost and organic matter to your garden to maintain and improve soil health. Compost enriches the soil with essential nutrients and improves its structure, making it more fertile and capable of retaining moisture. Organic matter, such as decomposed leaves and plant residues, provides a continuous food source for beneficial microorganisms, keeping the soil ecosystem vibrant. Avoiding soil compaction is another essential practice. Compacted soil restricts root growth and reduces air and water infiltration, creating conditions that favor disease development. Avoid walking on garden beds to prevent compaction and use pathways or stepping stones. Rotating crops is also vital for maintaining soil health. Crop rotation prevents the soil's buildup of pathogens and pests, reducing the risk of diseases. Planting different crops in each bed every season disrupts the life cycles of soil-borne diseases and pests, keeping your garden healthier.

Proper plant care techniques can significantly reduce the risk of diseases. Ensuring adequate spacing between plants is essential for good air circulation. Crowded plants create a humid environment that promotes the growth of fungal diseases. By spacing plants appropriately, you allow air to flow freely, reducing humidity and the likelihood of disease. Pruning is another effective technique. Regularly remove diseased or damaged parts of the plants to prevent the spread of pathogens. Pruning also improves air circulation and light penetration, creating a healthier environment for your plants. Watering at the base of plants, rather than overhead, helps avoid wet foliage, which can lead to fungal infections. Use a soaker hose or drip irrigation system to deliver water directly to the roots, keeping the leaves dry and reducing disease risk.

You can effectively manage common plant diseases with organic strategies. A milk spray or baking soda solution can control powdery mildew, a fungal disease that appears as a white, powdery coating on leaves. To make milk spray, mix one part of the milk with nine parts of water and spray it on the affected plants. The proteins in the milk help prevent the growth of mildew. Alternatively, mix one tablespoon of baking soda with a gallon of water and a few drops of liquid soap to create a baking soda solution. Spray this mixture on the plants to combat powdery mildew. Blight, a severe disease that affects tomatoes and potatoes, causes dark spots on leaves and stems, leading to plant death. To manage blight, promptly remove and destroy affected plant parts to prevent the disease from spreading. Improving air circulation by spacing plants properly and avoiding overhead watering can also help reduce blight incidence.

Root rot is another common disease caused by over-watering and poor drainage. It results in wilting, yellowing leaves and mushy roots. To prevent root rot, ensure proper drainage in your garden beds and containers. Use well-drained soil mixes and avoid over-watering. If you suspect root rot, remove the affected plant from the soil, trim off the diseased roots, and replant it in fresh, well-draining soil. These organic management strategies can help you maintain a healthy, disease-free garden.

Focusing on soil health and adopting proper plant care techniques can significantly reduce the risk of plant diseases. Promoting beneficial microorganisms, enhancing nutrient uptake, and maintaining a healthy soil structure create a resilient garden environment. Regularly adding compost and organic matter, avoiding soil compaction, and rotating crops are essential. Proper spacing, pruning, and watering techniques protect your plants from diseases. Organic management strategies, such as using milk spray for powdery mildew, removing affected parts for blight, and ensuring proper drainage to prevent root rot, effectively maintaining plant health. You can enjoy a thriving, disease-free garden with a bit of attention and care.

6.5 Companion Planting for Pest and Disease Management

Companion planting is a fascinating and effective strategy for managing pests and diseases in your garden. The concept revolves around growing different plants close together to enhance their growth and health while naturally deterring pests. This method creates a more resilient garden environment and reduces the need for chemical interventions. By carefully selecting plant pairs, you can use their natural properties to protect and nourish each other.

One of the primary benefits of companion planting is natural pest deterrence. Certain plants produce scents or chemicals that repel pests, keeping them away from more vulnerable neighbors. For instance, marigolds are well-known for repelling nematodes, tiny worms that can cause significant damage to plant roots. Planting marigolds alongside tomatoes can help keep these pests at bay, allowing your tomatoes to thrive. Similarly, basil planted with peppers can deter aphids while improving the peppers' flavor. The strong scent of basil confuses aphids and other pests, making it harder for them to locate their preferred plants. Nasturtiums, another excellent companion, attract aphids away from crops like cucumbers. This sacrificial strategy protects your main crops by drawing pests to the nasturtiums, which are more resilient and can handle the damage.

Enhanced plant growth and health are other significant advantages of companion planting. Some plants release substances into the soil that benefit their neighbors. For example, beans are legumes that fix nitrogen in the soil, enriching it for other plants. Growing beans with corn or squash can give these plants essential nutrients, boosting their growth. Additionally, the physical structure of some plants can offer support or shade to others. Tall plants like sunflowers can provide a natural trellis for climbing beans, while leafy greens like lettuce can benefit from the shade provided by taller plants, reducing the risk of bolting in hot weather.

The layout of your land is essential to maximize the effectiveness of companion planting. Inter-planting rows of companion plants can create a diverse and balanced ecosystem. For example, alternating rows of carrots and onions can deter pests that are attracted to the scent of one but repelled by the other. Creating mixed plant beds rather than monocultures can also help confuse pests and reduce the spread of diseases. Mixing different plant types in a single bed can create a more complex environment that is harder for pests to navigate. Using companion plants as borders or buffers is another effective strategy. Planting aromatic herbs like rosemary or sage around the edges of your garden can create a barrier that deters pests from entering.

Monitoring the effectiveness of your companion planting efforts is crucial. Regularly observe plant health and pest activity to determine if your strategies work. Adjust your approach if you notice that specific pairings must provide the expected benefits. Seasonal rotation of companion plants helps maintain their effectiveness. This practice prevents pests from becoming accustomed to the plant combinations and reduces the risk of soil depletion. Keeping a garden journal to record successful pairings and observations about plant health and pest activity can be incredibly useful. This information can guide your planting decisions in future seasons, helping you refine and improve your companion planting strategies.

Companion planting is a powerful tool in organic pest and disease management, offering natural deterrence and enhanced plant health. You can create a more resilient and productive garden by carefully selecting and arranging your plants. Regular monitoring and adjustments ensure that your companion planting efforts remain effective, leading to a thriving and balanced garden ecosystem.

In the next chapter, we will explore advanced soil management techniques that will further enhance the health and productivity of your urban garden. From no-till gardening to understanding soil microbiology, these methods will provide you with the knowledge to create an optimal growing environment.

Chapter Seven

Year-Round Urban Gardening

I remember the first time I harvested fresh lettuce in the middle of winter. It felt like a minor miracle. Living in a city, I had resigned to the idea that gardening was a seasonal activity limited by the whims of the weather. But that year, I discovered the power of season extenders. I transformed my urban garden into a year-round haven of fresh produce using simple tools like cold frames and greenhouses. This chapter will explore how you can extend your growing season and enjoy the fruits of your labor all year.

7.1 Extending the Growing Season: Cold Frames and Greenhouses

Cold frames and greenhouses are invaluable tools for any urban gardener looking to extend the growing season. A cold frame is a bottomless box placed over plants to protect them from cold weather. It has a transparent roof that lets in light while keeping out frost and strong winds. This simple structure can significantly extend your gardening season by creating a warmer and more stable microclimate than the surrounding environment. Cold frames are particularly useful in autumn and spring, allowing you to start your garden earlier and keep it going longer.

An image showing cold frames and greenhouses on a raised bed garden

Greenhouses, varying in type and size, cater to the specific needs of your plants by ensuring a controlled environment. Mini greenhouses are compact and portable, ideal for small balconies or patios, allowing urban gardeners to maximize their limited space efficiently. For those with a bit more room, portable greenhouses offer a flexible solution; they are easy to relocate and perfect for a backyard or larger outdoor areas. On the other end of the spectrum, permanent greenhouses represent the most substantial investment, providing a dedicated, year-round gardening space with the most stable conditions for plant growth.

Building a cold frame is a straightforward and rewarding project. You will need some basic materials:

1. Wood is used for the frame, an old window or clear plastic for the top, and hinges allow the top to open and close.

2. Start by measuring and cutting the wood to create a rectangular frame. The dimensions can vary based on the size of your space and the window or plastic sheet you have.

3. Assemble the frame using screws or nails, ensuring it is sturdy and square.

4. Using hinges, attach the window or plastic sheet to the top of the frame.

This addition allows you to open the top for ventilation on warmer days. Position your cold frame in a sunny spot to maximize the light and warmth it receives. The south side of your home or garden is often the best location.

Greenhouses, whether mini, portable, or permanent, require more planning and setup to be used effectively. Controlling temperature and humidity is crucial for maintaining a healthy, growing environment. During the day, the greenhouse can become quite warm, even in cooler weather. Ventilation is essential to prevent overheating and the buildup of mold and mildew. Many greenhouses come with built-in vents or windows that you can open to allow for airflow. You can also use fans to improve air circulation. At night, temperatures can drop significantly. Using grow lights can provide additional warmth and light, helping to maintain a stable environment for your plants. These lights mimic natural sunlight, promoting healthy growth despite short days.

Choosing suitable crops is critical to success when planting in cold frames and greenhouses. Cool season crops like lettuce, spinach, and radishes are ideal for cold frames. These plants thrive in cooler temperatures and can handle the lower light levels of early spring and late autumn. They are also relatively quick growing, allowing you to harvest and replant multiple times throughout the season. For greenhouses, you can start warm-season transplants like tomatoes, peppers, and cucumbers early, giving them a head start before transplanting them outside. This practice can lead to earlier and more abundant harvests. Greenhouses can also grow heat-loving plants that might struggle in cooler climates in the summer heat.

Cold Frame and Greenhouse Planting Tips

- **Cool-Season Crops:** Lettuces, spinach, radishes, kale, and peas thrive in the cooler temperatures of cold frames.

- **Warm-Season Transplants:** Start tomatoes, peppers, and cucumbers in greenhouses for a head start on the growing season.

- **Succession Planting:** Plant multiple rounds of quick-growing crops like lettuce and radishes to maximize your harvest.

- **Ventilation:** Open cold frames and greenhouse vents on warm days to prevent overheating and mold growth.

- **Grow Lights:** Use grow lights in greenhouses to provide additional warmth and light during shorter days.

Incorporating cold frames and greenhouses into your urban garden can revolutionize the way you grow. These season extenders allow you to enjoy fresh produce year-round, making the most of your space and resources. Whether building a simple cold frame or investing in a greenhouse, the benefits are well worth the effort.

7.2 Indoor Gardening: Techniques and Tips for Year-Round Growth

Indoor gardening has revolutionized how we grow plants, especially in urban settings with limited outdoor space. One of the most significant advantages of indoor gardening is protection from extreme weather conditions. Whether it is a scorching summer day or a freezing winter night, indoor gardens remain unaffected by the elements. This stability allows for a more consistent growing environment, free from the unpredictable swings of nature. Additionally, indoor gardening offers unparalleled control over the growing environment. You can adjust factors like light, temperature, and humidity to suit the needs of your plants, creating an ideal setting for growth. The convenience of having your garden within arm's reach cannot be overstated. You can tend to your plants without worrying about the weather or pests. This accessibility makes indoor gardening a practical option for busy urban dwellers who want to incorporate more greenery into their lives.

Choosing the right indoor gardening system is crucial for success. Hydroponics, for instance, involves growing plants in a nutrient solution rather than soil. This highly efficient method can result in faster growth and higher yields. Hydroponic systems come in various forms, from simple setups involving a nutrient solution and a water pump to more complex systems with automated feeding and lighting controls. Aeroponics takes this further by growing plants in an air/mist environment. The plant roots suspend in the air and get misted with a nutrient solution, allowing maximum oxygen exposure and nutrient absorption. This method is suitable for growing leafy greens and herbs. Traditional soil-based containers are another option. These involve using pots and planters filled with soil, much like an outdoor garden but scaled down for indoor use. They are easy to manage and a good starting point for beginners. Each system has pros and cons; the best choice depends on your needs, space, and experience level.

Lighting is a critical component of indoor gardening. Sunlight may not be enough, especially in cities where buildings and weather can block it. Grow lights step in to fill this gap. Fluorescent lights are a popular choice for their affordability and efficiency. They are suitable for growing seedlings and leafy greens. LED lights are more energy-efficient and have a longer lifespan. Indoor gardeners can customize them to emit specific light spectrums, making them ideal for different stages of plant growth. High-Intensity Discharge (HID) lights are another option, providing intense light that supports robust growth, especially for flowering plants. The duration and intensity of light are crucial. Most plants require 12–16 hours of light daily for optimal growth. Proper positioning of the light devices ensures even coverage. Place the lights close to the plants but ensure they do not cause overheating or burning. Adjusting the height as the plants grow ensures they receive consistent light.

Maintaining an indoor garden involves regular care and attention. Managing indoor pests is a common challenge. Spider mites and fungus gnats are frequent culprits. Regularly inspecting your plants and using natural remedies like neem oil or insecticidal soap can help keep these pests in check. Watering schedules are also important. Indoor plants can dry out quickly due to the controlled environment, so regular watering is essential. However, over-watering can lead to root rot. Using a moisture meter can help you find the right balance. Fertilization is another crucial aspect. Indoor plants may require frequent feeding since they do not draw nutrients from an outdoor soil ecosystem. Using a balanced, water-soluble fertilizer ensures your plants get the necessary nutrients. Monitoring humidity levels with a hygrometer is also beneficial. Most indoor plants thrive in humidity levels between 40-60%. If the air is too dry, especially in winter, using a humidifier can help maintain the ideal conditions.

Indoor gardening offers a unique opportunity to grow plants year-round, regardless of the weather outside. You can create a thriving indoor garden by choosing the right system, providing adequate lighting, and maintaining regular care. The benefits of indoor gardening extend beyond just growing plants; it is about creating a green space that enhances your living environment and brings a touch of nature into your home.

7.3 Selecting Plants for Different Seasons

Choosing the right plants for each season ensures continuous growth and harvest in your urban garden. The key lies in understanding the specific needs of cool-season and warm-season crops and hardy crops that can withstand winter's chill. Cool-season crops are ideal for spring and fall. They thrive in the lower temperatures of these seasons and can often handle light frosts. Vegetables like spinach, kale, and arugula are perfect examples. These leafy greens grow quickly and provide a nutritious addition to your diet. Root vegetables, such as carrots, beets, and radishes, also fall into this category. You can sow them early in the spring and again in the fall, giving you two harvests in one year. Their ability to mature before the heat of summer or the deep cold of winter makes them versatile and reliable choices.

Warm-season crops come into their own during the summer months. These plants require higher soil and air temperatures to grow and produce fruit. Tomatoes, peppers, and cucumbers are classic examples. These fruit-bearing plants need plenty of sunlight and warmth to thrive. Caring for them involves:

- Regular watering.

- Mulching to retain soil moisture.

- Staking or trellising to support their growth.

Herbs like basil, oregano, and thyme also flourish in the summer. They add flavor to your meals and attract beneficial insects to your garden. Planting these herbs in containers or with vegetables can create a harmonious and productive garden space.

As temperatures drop in the fall, turning your attention to hardy crops can keep your garden productive. Brassicas like Broccoli, cauliflower, and Brussels sprouts are excellent choices. These plants exhibit resilience to cooler temperatures and can even enhance flavor after a light frost. Hardy greens like collards, chards, and mustard greens also thrive in the cooler months. These greens can be continually harvested, providing fresh leaves throughout the fall and early winter. They are particularly well-suited for raised beds and containers, where they can be easily protected from early frosts with row covers or cloches.

Focusing on leafy greens and root vegetables sets the stage for a successful gardening year in spring. Spinach, kale, and arugula are among the first crops you can plant. These greens quickly germinate and grow, allowing you to start harvesting within a few weeks. When the soil is ready for planting, you can sow carrots, beets, and radishes directly into it. They grow quickly and can often be harvested before the heat of summer sets in. Spring is also an excellent time to plant peas and onions, which thrive in the cooler weather and provide a steady supply of fresh produce.

Summer is the season for warm-season crops like tomatoes, peppers, and cucumbers. These plants require consistent care to ensure a bountiful harvest. Regular watering is essential, as the heat can quickly dry the soil. Mulching helps retain moisture and keep the root zone cool. Staking or using cages for tomatoes and trellises for cucumbers can support their growth and make harvesting easier. Basil, oregano, and thyme can be planted alongside these vegetables, benefiting from the same care and adding culinary and ecological value to your garden.

As fall approaches, turning your attention to brassicas and hardy greens ensures your garden remains productive. You can plant Broccoli, cauliflower, and Brussels sprouts in late summer to mature in the cooler fall weather. These plants are cold-tolerant and can withstand light frost. You can sew collards, chards, and mustard greens in succession to provide a continuous harvest. These greens are nutritious and resilient, making them reliable for fall and early winter gardening.

Incorporating a variety of plants suited to different seasons ensures that your urban garden remains productive and vibrant throughout the year. You can enjoy a continuous supply of fresh produce by understanding and catering to the specific needs of cool-season, warm-season, and hardy crops. This approach maximizes your garden's potential and adds diversity and resilience to your urban space.

7.4 Winter Gardening: Strategies for Cold Weather

Winter gardening presents unique challenges that require thoughtful strategies. Reduced daylight hours can significantly impact plant growth, as many vegetables need ample light to thrive. The shortened days mean less photosynthesis, stunting growth, and delaying harvests. Cold temperatures and frost pose another challenge. Frost can damage or kill delicate plants, while prolonged cold spells can freeze the soil, making it difficult for roots to access water and nutrients. Limited plant growth is also a concern, as the cold slows down metabolic processes in plants, resulting in slower development and smaller yields.

Several methods can protect your plants from harsh winter weather. Row covers and cloches are simple yet effective tools. Row covers are lightweight fabrics that can be draped over plants to shield them from frost and wind while allowing light and moisture to pass through. Cloches are small, dome-shaped covers that provide more localized protection and can be made from materials like plastic or glass. Both methods create a microclimate around the plants, raising the temperature slightly and reducing the risk of frost damage. Mulching is another valuable technique. A thick layer of organic mulch, such as straw or leaves, insulates the soil and helps retain moisture. This insulation keeps the soil temperature stable, protecting plant roots from freezing temperatures. For potted plants, bringing them indoors can be a lifesaver. Placing them in a sunny window or under grow lights can provide the warmth and light they need to survive the winter months.

Selecting cold-hardy plants is essential for a successful winter garden. Perennials like Kale and Brussels sprouts are excellent choices. These vegetables are resilient to cold and can improve flavor after frost. Kale becomes sweeter as the cold converts its starches into sugars. Brussels sprouts also benefit from the cold, becoming more tender and flavorful. Root vegetables like parsnips and turnips are well-suited for winter gardening. Farmers can leave these crops in the ground even after the first frost and harvest them throughout the winter. The cold ground is a natural refrigerator, preserving these root vegetables and enhancing their sweetness.

Effective Management of winter gardens involves a few essential techniques. Adjusting watering schedules is crucial to prevent the soil from freezing. Watering in the morning allows excess water to evaporate during the day, reducing the risk of ice formation at night. Using reflective surfaces can maximize the available light, which is particularly helpful during the short winter days. Mirrors or white-painted surfaces placed around the garden can reflect sunlight onto the plants, increasing light exposure and promoting growth. Harvesting winter crops requires a gentle touch. Cold weather can make plants more brittle, so using a sharp knife or garden shears can help avoid damaging the plants. Additionally, harvesting during the warmest part of the day, when the plants are less rigid, can make the process easier.

These winter gardening strategies can help you overcome the challenges of reduced daylight, cold temperatures, and limited growth. Protective methods like row covers, cloches, and mulching can shield your plants from the worst winter weather. Selecting cold-hardy plants ensures you have resilient crops that thrive despite the cold. Adjusting your watering schedules, using reflective surfaces, and employing careful harvesting techniques can further enhance your winter gardening success. These practices allow you to maintain a productive garden even in the depths of winter, providing fresh, homegrown produce year-round.

7.5 Maintaining Soil Health Through Seasonal Changes

Maintaining soil health throughout the seasons is crucial for the success of your year-round urban garden. Every season presents unique challenges and opportunities for soil management. Knowing how to handle soil during these changes can prevent nutrient loss and soil compaction and promote beneficial organisms. Seasonal maintenance is about adding fertilizers and creating a robust ecosystem that supports plant growth throughout the year.

Adding compost to your soil in the spring provides a much-needed nutrient boost after the long winter months. Compost enriches the soil with organic matter and essential nutrients, preparing it for the upcoming growing season. This practice improves soil structure and increases its ability to retain water and nutrients. Green manure can further enhance soil health as the weather warms up in the summer. Green manure involves growing specific plants, such as clover, rye, and vetch, which are then tilled back into the soil. These plants add organic matter, improve soil structure, and provide nutrients as they decompose. In the fall, applying mulch helps protect the soil from the harsh winter ahead. Mulch acts as an insulator, maintaining soil temperature and moisture levels. It also prevents erosion and suppresses weed growth, making the soil healthier when spring returns.

Cover cropping is another powerful technique for maintaining soil health. Cover crops are planted to cover the soil rather than for harvest. They help prevent erosion, improve soil structure, and increase organic matter. Choosing the right cover crops, such as clover, rye, and vetch, is crucial. These plants grow quickly and can be easily incorporated into the soil. The timing of planting cover crops is also essential. They should be sown after your main crops are harvested and before the onset of winter. Once the cover crops have grown, they can be tilled into the soil, enriching it with nutrients and organic matter. This practice improves soil fertility and promotes a healthy soil ecosystem by supporting beneficial microorganisms.

Regularly monitoring and adjusting soil conditions is essential for maintaining soil health. Conducting soil tests for pH and nutrient levels provides valuable insight into what your soil needs. pH levels affect nutrient availability, and knowing whether your soil is too acidic or alkaline allows you to make necessary adjustments. Adding lime can raise the pH, while sulfur can lower it. These adjustments ensure that your plants can access the nutrients they need. Adjusting watering and fertilization practices according to the season is also essential. In the hot summer, soil may dry out more quickly, requiring frequent watering.

Conversely, overwatering can lead to root rot and other issues in the cooler months. Plant health can also serve as an indicator of soil quality. Yellowing leaves, stunted growth, or poor yields may signal nutrient deficiencies or other soil problems. Addressing these issues promptly by amending the soil or adjusting your care practices can keep your garden thriving.

Maintaining soil health through seasonal changes involves a combination of regular amendments, cover cropping, and careful monitoring. Adding compost in the spring, using green manure in the summer, and mulching in the fall can give your soil the nutrients and protection it needs. Cover cropping further enhances soil health by adding organic matter and supporting beneficial microorganisms. Regular soil testing and observation allow you to make informed adjustments, ensuring your soil remains fertile and productive. These practices create a resilient soil ecosystem that supports year-round gardening success.

Understanding and implementing these seasonal soil maintenance practices can create a thriving urban garden that produces healthy, vibrant plants throughout the year. Maintaining soil health is the foundation of sustainable gardens, whether growing vegetables, herbs, or flowers. These strategies enhance your garden's productivity and contribute to environmental sustainability by promoting soil conservation and reducing the need for chemical inputs.

Chapter Eight

Sustainable and Eco-Friendly Practices

O ne brisk morning, I stood in front of a neglected garden plot. The soil was compacted, the plants struggled, and I felt overwhelmed. Then, I stumbled upon the concept of permaculture philosophy that promised not just a garden but a thriving, self-sustaining ecosystem. This chapter focuses on permaculture principles and guides you in integrating them into your urban garden, transforming even the most neglected spaces into vibrant, productive havens.

8.1 Principles of Permaculture in Urban Gardens

Permaculture is grounded in three foundational principles: care for the earth, care for people, and fair share. These principles guide every decision you make in your garden, ensuring that your practices are sustainable and beneficial for the environment and community.

Care for the earth means prioritizing the environment's health in all your gardening activities. This practice involves minimizing waste, conserving resources, and enhancing biodiversity. For instance, you might opt for compost made from kitchen scraps instead of chemical fertilizers. Doing so reduces waste and enriches the soil with organic matter, promoting a healthier ecosystem.

Care for People emphasizes the importance of creating gardens that meet human needs while fostering community well-being. This activity could mean designing spaces that are accessible to everyone or growing various plants to ensure a steady supply of fresh produce. In an urban setting, this might translate to community gardens where neighbors can gather, share knowledge, and support each other.

Fair share involves redistributing surplus to ensure that resources are used efficiently and equitably. This principle encourages sharing excess produce, seeds, or knowledge with others. This practice could mean donating extra vegetables to a local food bank or teaching neighbors to start their garden.

Design strategies for urban permaculture are tailored to make the most of limited space while adhering to these principles. Creating closed-loop systems is a crucial strategy. Recycling waste and composting organic matter can create a self-sustaining cycle that reduces the need for external input. For example, you can compost your kitchen scraps to enrich the soil, which supports plant growth. The plants produce organic matter that can be composted again, creating a continuous sustainability loop.

Another effective strategy is to use vertical space for multi-layered planting. Growing plants vertically maximize your garden's productivity without additional ground space. This practice could involve trellises for climbing plants, hanging baskets, or even green walls. Vertical gardens are space-efficient, add aesthetic appeal, and help improve air quality.

Incorporating water catchment systems is essential for conserving water and ensuring a steady supply for your garden. Simple methods like installing rain barrels to collect runoff from roofs can significantly reduce your reliance on municipal water. More advanced systems might include Swales—shallow, water-retentive ditches designed to capture and slowly release water into the soil. These systems help reduce runoff, prevent erosion, and ensure that plants can access water even during dry periods.

Integrating permaculture with urban infrastructure involves finding creative ways to blend natural elements with artificial structures. Green roofs and walls are excellent examples. These installations provide insulation, reduce energy costs, and create habitats for pollinators and other beneficial organisms. A green roof can transform a barren rooftop into a lush garden space, while green walls can add beauty and function to vertical surfaces.

Rainwater harvesting is another practical application. By capturing and storing rainwater, you can reduce your water bill and ensure your garden remains hydrated. Simple systems like rain barrels can be easily installed and maintained, while more complex setups might involve underground cisterns or above-ground tanks. The captured water can be used for irrigation, reducing your dependence on treated water and promoting sustainability.

Community gardening spaces embody permaculture principles by fostering collaboration and resource sharing. These gardens can transform unused urban areas into productive, green spaces that benefit the entire community. Participants share tools, knowledge, and produce, creating a sense of camaraderie and mutual support. Community gardens also provide educational opportunities, teaching participants about sustainable practices and the importance of environmental stewardship.

Real-world examples of urban permaculture projects highlight the transformative power of these principles. The 18th and Rhode Island Permaculture Garden in San Francisco is a striking testament to what can be achieved. This once-empty city lot has been transformed into a productive garden that produces thousands of pounds of food. The garden's design includes swales to capture water, reducing runoff and erosion, and perennial plants arranged in guilds to support each other's growth. Nutrients and organic matter are maintained on-site in a closed-loop system, ensuring the soil remains productive.

A permaculture-inspired community garden in Berlin has brought neighbors together to create a thriving green space. This garden incorporates raised beds, composting systems, and water catchment methods to create a self-sustaining environment. The community shares the harvest, ensuring everyone benefits from the collective effort. This project provides fresh produce, strengthens community bonds, and promotes sustainable living.

These examples show that permaculture principles can be successfully applied in urban settings, transforming even the smallest spaces into productive, sustainable gardens. By incorporating these principles into your garden, you can create a space that is not only beautiful and productive but also environmentally responsible and community-oriented. Whether you are working with a small balcony, a rooftop, or a community plot, permaculture offers practical solutions for sustainable urban gardening.

8.2 Water Conservation Techniques for Urban Gardens

Water conservation is critical to urban gardening, especially considering many cities' environmental impact and resource limitations. Conserving water means reducing waste and ensuring that your garden remains healthy and productive over the long term. It is about saving money on your water bill and creating a sustainable practice that benefits your garden and the planet. Efficient water use helps maintain soil health by preventing nutrient leaching and reducing soil erosion, which are common problems in over-watered gardens.

Drip irrigation systems are one of the most effective techniques for efficiently watering plants. These systems deliver water directly to the plant roots through a network of tubes and emitters, minimizing evaporation and runoff—drip irrigation benefits container gardens and raised beds, where precise water control is necessary. By placing emitters at the base of each plant, you ensure that water reaches where it's needed most, reducing waste and promoting healthier plants. Another advantage of drip irrigation is that it can be easily automated with timers, making it ideal for busy urban gardeners who might not have time to water their plants manually every day.

Soaker hoses offer another efficient watering solution. Made from porous material, these hoses release water slowly along their entire length, ensuring even distribution. They are beneficial for long rows of plants or more extensive garden beds. By laying soaker hoses along the base of your plants and covering them with mulch, you can keep the soil consistently moist without over-watering. This method reduces surface evaporation and encourages deep root growth, making plants more resilient to drought conditions.

Hand watering with precision tools can be highly effective for smaller gardens or specific plants that need extra attention. Using a watering can with a fine nozzle or a hose with a gentle spray attachment allows you to target individual plants without wasting water. This method benefits young seedlings or delicate plants that require careful handling. By observing your plants closely as you water them, you can adjust your watering practices based on their specific needs, ensuring that each plant receives just the right amount of moisture.

Rainwater harvesting systems provide an excellent way to capture and store rainwater for your garden. Installing rain barrels at your home's downspouts is simple and effective. These barrels collect runoff from your roof, which can then water your plants. By using rainwater, you can reduce your reliance on municipal water supplies and save money. You can implement more complex systems like rain gardens or Swales for more extensive gardens. Rain gardens capture and filter rainwater, allowing it to percolate into the ground, reducing runoff and helping recharge groundwater supplies. Swales are shallow, water-retentive ditches that direct water to specific areas of your garden, preventing erosion and ensuring that water is available where it's needed most.

Gray water recycling is another innovative way to conserve water. Grey water refers to lightly used water from laundry, showers, and sinks (excluding kitchen sinks because of grease and food particles). Setting up a gray water system involves diverting this water to your garden, where you can irrigate plants. Using environmentally friendly soaps and detergents is essential to avoid harming your plants. Simple systems can involve manually collecting gray water with buckets, while more advanced setups might include plumbing modifications to direct gray water to specific areas of your garden. You must follow safety considerations and local regulations to ensure that gray water is used responsibly and does not contaminate your garden or the surrounding environment.

Water conservation in urban gardens is a practical necessity and a way to contribute positively to environmental sustainability. By implementing efficient watering techniques like drip irrigation, soaker hoses, and hand watering, you can ensure that your plants receive the moisture they need without wasting water. Rainwater harvesting and gray water recycling further enhance your garden's sustainability, reducing your reliance on municipal water supplies and making the most available resources. These practices not only support the health of your plants but also promote a more sustainable and environmentally conscious approach to gardening.

8.3 Using Organic Fertilizers: Types and Applications

Enhancing your soil with organic fertilizers is one of the best ways to ensure a thriving urban garden. Unlike synthetic fertilizers, which can degrade soil health over time, organic fertilizers improve soil structure and promote beneficial organisms. When you add organic matter to your soil, you feed your plants and support the entire ecosystem. Healthy soil teems with life—bacteria, fungi, earthworms, and other organisms that break down organic matter and release nutrients in a form that plants can absorb. This dynamic system creates a fertile environment where plants can thrive and resist pests and diseases.

Various organic fertilizers are suitable for urban gardens, each offering unique benefits. Compost is the most versatile. Made from decomposed organic material, it enriches the soil with essential nutrients and improves its structure. You can also create compost tea by steeping compost in water, which provides a quick nutrient boost for your plants. Worm castings, another excellent option, are the byproduct of vermiculture. These castings are rich in nutrients and beneficial microorganisms, making them a potent soil amendment. Manure and guano are traditional organic fertilizers that add nutrients to the soil. They must be well-aged or composted to avoid burning plants with high nitrogen content. Bone and blood meals are high in phosphorus and nitrogen, respectively, and are great for promoting root development and overall plant growth.

Applying organic fertilizers involves several techniques to ensure your plants benefit most. Before planting, mix the organic fertilizers into the soil. This practice allows the nutrients to integrate well with the soil structure, providing a balanced nutrient supply as the plants grow. During the growing season, you can top-dress the soil with compost or worm castings. Spread a thin layer around the base of your plants and gently work it into the top layer of soil. This method provides a slow-release source of nutrients that feeds your plants. Compost tea can be applied as a liquid fertilizer. To make it, fill a container with water, add a bag of compost, and let it steep for 24-48 hours. Strain the liquid and use it to water your plants, giving them an immediate nutrient boost and enhancing microbial activity in the soil.

DIY organic fertilizers are a cost-effective way to nourish your garden using materials you likely already have at home. Banana peel tea is an excellent source of potassium, which strengthens plant stems and roots. To make it, soak banana peels in water for a few days, then use the resulting liquid to water your plants. Eggshell powder is another easy-to-make fertilizer that provides calcium essential for strong cell walls and root development. Collect and clean eggshells, dry them and grind them into a fine powder. Sprinkle the powder around the base of your plants or mix it into the soil. Fish emulsion from kitchen scraps like fish bones and heads is rich in nitrogen and other essential nutrients. To make it, place the fish scraps in a bucket, cover it with water, and let it ferment for a few weeks. Strain the liquid and dilute it with water before applying it to your garden.

Organic fertilizers benefit plants and contribute to a healthier, more sustainable environment. Unlike synthetic fertilizers, which can lead to nutrient runoff and pollution, organic fertilizers improve the soil's ability to retain water and nutrients. They also encourage the growth of beneficial microorganisms that help plants absorb these nutrients more effectively. By opting for organic fertilizers, you are investing in the long-term health of your garden and the broader ecosystem.

Incorporating these organic fertilizers into your gardening routine can transform your urban garden into a productive, vibrant space. Whether using compost, worm castings, or DIY fertilizers, each application enriches your soil and supports robust plant growth. By understanding the benefits and techniques of using organic fertilizers, you can create a thriving garden that is both beautiful and sustainable.

8.4 Mulching: Benefits and Methods for Urban Gardens

Mulching is a simple yet powerful technique that offers multiple benefits for urban gardens. One of the most significant advantages of mulching is its ability to retain soil moisture. Covering the soil with a layer of mulch reduces evaporation, ensuring that the water stays in the soil longer. This practice is essential in urban environments, where the heat from buildings and pavements can quickly dry out the soil. Mulching is a protective barrier, keeping the soil cool and moist, even during hot summer days. Another critical benefit is weed suppression. Weeds compete with your plants for nutrients, water, and sunlight. A thick layer of mulch prevents weed seeds from germinating and reaching the surface, reducing the need for manual weeding and chemical herbicides.

Regulating soil temperature is another crucial function of mulching. In the summer, mulch keeps the soil cool by blocking direct sunlight. In the winter, it acts as an insulating layer, protecting plant roots from freezing temperatures. This temperature regulation helps create a stable environment for your plants, promoting consistent growth and reducing stress. As mulch decomposes, it adds organic matter to the soil, enriching it with nutrients and improving its structure. This process enhances soil fertility, making your garden more productive and resilient. The decomposed organic matter also encourages beneficial microorganisms, which are vital in nutrient cycling and soil health.

There are various types of mulch, each suited to different gardening needs. Organic mulches, such as straw, wood chips, and leaves, are popular. Straw is lightweight and easy to spread, making it ideal for vegetable gardens. Wood chips are excellent for flower beds and around shrubs, providing long-lasting coverage. Leaves, readily available in the fall, can be shredded and used as mulch, adding valuable organic matter to the soil. Inorganic mulches, including gravel, stones, and plastic sheeting, are also options. Gravel and stones provide a clean, finished look for pathways and decorative areas. Plastic sheeting is often used in commercial agriculture to suppress weeds and conserve moisture, but it can be less suitable for home gardens because of its environmental impact. Living mulches offer another option, such as cover crops and ground covers. These plants grow close to the ground, providing continuous coverage and adding organic matter as they decompose. Clover and creeping thyme are examples of living mulches that can enhance soil health while reducing erosion.

Proper mulching techniques are essential to maximizing the benefits. When applying mulch, aim for a thickness of 2 to 4 inches. This depth can suppress weeds and retain moisture without smothering the plants. Spread the mulch evenly, ensuring complete coverage while leaving a small gap around the base of each plant to prevent rot and disease. Seasonal adjustments may be necessary. In the spring, remove some mulch to allow the soil to warm up. In the fall, adding a fresh layer of mulch can help insulate the soil and protect plant roots during the winter months.

Creative mulching ideas can enhance both the aesthetics and functionality of your garden. Using recycled cardboard or newspaper as a base layer under organic mulch effectively suppresses weeds while recycling materials that might otherwise end up in a landfill. Lay down the cardboard or newspaper, soak it in water, and cover it with your mulch. For acid-loving plants like blueberries and azaleas, mulching with pine needles can help maintain the acidic soil conditions they prefer. Decorative mulch options, such as colored wood chips and pebbles, can add visual interest to your garden, creating a professional look. Colored wood chips come in various shades, allowing you to match the mulch to your garden's color scheme. Pebbles and small stones can create patterns or borders, adding texture and contrast.

Mulching is a versatile and practical practice that can significantly improve the health and productivity of your urban garden. The benefits are clear whether you use organic materials like straw and leaves, inorganic options like gravel, or living mulches like cover crops. Mulching supports a thriving, sustainable garden by retaining moisture, suppressing weeds, regulating temperature, and adding organic matter. Proper application techniques and creative ideas can further enhance your garden's appearance and functionality, making mulching an indispensable part of your gardening routine.

8.5 DIY Gardening Projects: Upcycling for Eco-Friendly Gardens

Upcycling is repurposing discarded materials into something valuable and functional. In urban gardening, upcycling reduces waste, encourages creativity, and provides cost-effective solutions. Instead of buying new garden supplies, you can transform everyday items into helpful gardening tools and decor. This approach aligns perfectly with eco-friendly practices, allowing you to create a unique and sustainable garden space.

One of the simplest ways to upcycle is by creating planters from old pallets and crates. Businesses often discard these materials, which you can find for free or cheaply. To make a pallet planter, first, choose a sturdy pallet. Sand it down to remove rough edges, then line the inside with landscape fabric to hold the soil. Fill it with a good soil mix and plant your favorite herbs or flowers.

Similarly, you can use crates. Drill holes at the bottom for drainage, line them with landscape fabric, and fill them with soil. These planters are perfect for small spaces and can be stacked or hung on walls to maximize vertical space.

Building raised beds from recycled wood is another practical upcycling project. Look for old wooden planks, doors, or pallets you can repurpose. Measure and cut the wood to the desired size, then assemble the pieces into a rectangular frame. Secure the corners with screws or brackets. Place the frame on your chosen spot, fill it with a rich soil mix, and start planting. Raised beds improve soil quality and drainage, making them ideal for urban gardening. Plus, using recycled wood adds a rustic charm to your garden.

You can make garden trellises from discarded materials like old ladders, bed frames, or bicycle wheels. These structures support climbing plants and add a vertical element to your garden. To create a trellis from an old ladder:

1. Stand it upright and secure it in the soil. Plant climbing vegetables like beans or cucumbers at the base, which will naturally grow up the rungs.

2. For a more artistic approach, use bicycle wheels.

3. Attach them to a sturdy frame or post and weave strings or wires between the spokes so the plants can climb.

This practice supports your plants and adds a whimsical touch to your garden.

Upcycling extends to garden decor as well. Crafting garden art from metal scraps is a fantastic way to add personality to your space. Collect pieces of scrap metal, like old tools, hardware, or kitchen utensils. Use a welder or strong adhesive to assemble them into sculptures or decorative stakes. These unique art pieces can be placed among your plants, adding a creative element to your garden. Designing pathways with broken tiles is another great idea. Instead of throwing away chipped or cracked tiles, use them to create mosaic pathways. Lay the tiles in a pattern along your garden paths, filling the gaps with sand or gravel. This practice recycles the tiles and creates beautiful and functional walkways.

Bird feeders and houses made from plastic bottles are both practical and eco-friendly. To make a bird feeder, cut openings on the sides of a plastic bottle, leaving space for the birds to perch. Fill the bottle with birdseed and hang it from a tree or a hook. For a birdhouse, cut an entrance hole at the front of the bottle and add a small perch below it.

Decorate the bottle with paint or stickers, then hang it in a sheltered spot. These projects provide shelter and food for local birds while reducing plastic waste.

Community upcycling initiatives have a significant impact on urban gardening. Many communities' gardens use upcycled materials to build raised beds, trellises, and compost bins. These projects save money and foster a sense of community and collaboration. Workshops and events promoting upcycling in gardening are becoming increasingly popular. These gatherings offer hands-on experience and teach participants how to turn everyday waste into garden treasures. They also raise awareness about sustainability and encourage more people to adopt eco-friendly practices.

In conclusion, upcycling offers endless possibilities for creating a sustainable and visually appealing urban garden. Repurposing materials reduces waste, saves money, and adds unique, creative elements to your space. Whether you build planters from pallets, craft garden art from metal scraps, or take part in community upcycling projects, each effort contributes to a greener, more sustainable world. Embrace the creativity and resourcefulness that upcycling brings to your gardening journey, and watch your urban garden thrive unexpectedly.

Chapter Nine

Advanced Soil Management Techniques

O ne chilly autumn day, I stood in my garden, a shovel in hand, staring at the compacted, lifeless soil. The thought of turning it over yet again felt exhausting. Then, I stumbled upon a concept that changed my approach to gardening entirely: no-till gardening. This method promised to preserve soil health without the back-breaking labor of traditional tilling. Intrigued, I decided to try it, and the results were nothing short of transformative.

9.1 No-Till Gardening: Benefits and Techniques

No-till or no-dig gardening minimizes soil disturbance to maintain its structure and health. Unlike traditional tilling, which involves turning and breaking up the soil, no-till gardening focuses on keeping the soil intact. Traditional tilling can cause soil compaction, erosion, and loss of organic matter. In contrast, no-till gardening aims to preserve the soil's natural ecosystem, allowing beneficial organisms to thrive and improving overall soil health.

The benefits of no-till gardening are numerous and impactful. One of the most significant advantages is the preservation of soil structure. When you avoid disturbing the soil, its natural layers remain intact, promoting better water infiltration and root growth. This practice, in turn, prevents erosion, as the soil is less likely to be washed away by rain. Enhanced water retention is another crucial benefit. No-till soils tend to hold moisture more effectively, reducing the need for frequent watering and helping plants survive during dry spells.

No-till gardening fosters a diverse and thriving community of beneficial soil organisms. These microorganisms play a crucial role in breaking down organic matter, fixing nitrogen, and decomposing plant residue, all of which contribute to a healthier, more fertile soil. Moreover, no-till gardening reduces labor and time investment. With no heavy tilling, you can focus your efforts on other aspects of gardening, making it a more sustainable and enjoyable practice.

Implementing no-till gardening involves several techniques to create a healthy and productive garden. One of the foundational practices is using cover crops to protect and enrich the soil. Cover crops, such as clovers, rye, and vetch, are planted during the off-season to prevent erosion, improve soil structure, and add organic matter. As they grow, these cover crops create a protective layer that shields the soil from harsh weather conditions. When it is time to plant your main crops, the cover crops can be cut down and left on the soil surface as mulch, providing a continuous supply of nutrients.

Applying mulch is another essential technique in no-till gardening. Mulch serves multiple purposes: it suppresses weeds, retains moisture, and adds organic matter to the soil as it decomposes. Organic mulches, such as straw, wood chips, and leaves, are particularly effective. Spread a thick mulch over your garden beds, ensuring the soil remains covered throughout the growing season. This protective layer helps regulate soil temperature, reduces evaporation, and creates a favorable environment for beneficial organisms.

Direct planting into crop residues is a practical and efficient way to implement no-till gardening. Instead of removing the remnants of previous crops, please leave them in place and plant your new crops directly into the residues. This technique allows the organic matter to break down naturally, enriching the soil and providing a continuous source of nutrients. It also minimizes soil disturbance, preserving the soil structure and promoting the growth of beneficial microorganisms.

While no-till gardening offers many benefits, it also comes with its own set of challenges. One common challenge is managing weeds without tilling. Since tilling can bury weed seeds and disrupt their growth, no-till gardeners need alternative strategies. Mulching is one practical solution, as it suppresses weed growth by blocking sunlight. Hand-weeding and using tools like hoes and weed pullers can help keep weeds in check. Another challenge is adapting to initial soil conditions. If your soil is heavily compacted or needs more nutrients, it may take time for no-till practices to yield noticeable improvements. Patience is vital, as the benefits of no-till gardening become more apparent over time. Consider incorporating compost and organic fertilizers to boost soil fertility during the transition period.

By embracing no-till gardening, you can create a healthier, more sustainable garden that requires less effort and yields better results. This method preserves soil structure, enhances water retention, and promotes a thriving community of beneficial organisms. Techniques like using cover crops, applying mulch, and direct planting into crop residues work together to create a productive and resilient garden. While challenges like weed management and initial soil conditions may arise, gardeners can overcome them with patience and the right strategies. No-till gardening offers a holistic approach to soil management, allowing you to cultivate a thriving garden while minimizing environmental impact.

9.2 Hydroponics and Aeroponics: Soil-Less Urban Gardening

Hydroponic gardening is a method that allows you to grow plants without soil by using a nutrient-rich water solution. The basic principle is simple: plants receive all the essential nutrients they need directly through the water, bypassing the soil entirely. This method can be incredibly beneficial for urban gardeners, especially those with limited space. Hydroponics enables faster plant growth and higher yields due to the optimal nutrient availability and controlled environment. The absence of soil also means fewer pests and diseases, making it easier to maintain healthy plants.

Several types of hydroponic systems are suited to different plants and space constraints. The Nutrient Film Technique (NFT) involves a continuous flow of nutrient solution over the roots of the plants. This method is effective for plants with smaller root systems, such as leafy greens and herbs. The roots are suspended in a sloping channel, allowing the nutrient solution to flow over them and drain back into the reservoir. This system ensures the roots receive a constant supply of nutrients and oxygen.

Deep Water Culture (DWC) is another popular hydroponic system where the roots of the plants are submerged directly into the nutrient solution. An air pump provides oxygen to the roots through an air stone, ensuring they remain healthy and vibrant. This system is easy to set up and maintain, making it ideal for beginners. It works well for fast-growing plants like lettuce and spinach, which thrive in an oxygen-rich environment.

Drip systems and wick systems are also widely used in hydroponic gardening. In a drip system, the nutrient solution is pumped through a network of tubes and dripped directly onto the base of each plant. This method allows for precise control over the amount of nutrients each plant receives, making it suitable for various plants, including tomatoes and peppers. On the other hand, Wick systems are more straightforward and do not require pumps or electricity. Nylon wicks draw the nutrient solution from a reservoir to the plant roots. This system works well for small plants and herbs but may need more for more extensive, heavy-feeding plants.

Aeroponics takes the concept of soil-less gardening to the next level. In an aeroponic system, plants are suspended in the air, and their roots are misted with a nutrient-rich solution. This method provides the highest oxygenation level to the roots, promoting faster growth and higher yields. Aeroponics is incredibly efficient in using water and nutrients, as the fine mist ensures that the plants absorb what they need with no waste. This system is ideal for urban gardeners looking to maximize their space and resources.

Setting up a hydroponic or aeroponic system in an urban environment requires careful planning and the right components. First, choose the system that best fits your space and the plants you want to grow. For instance, an NFT or aeroponic system might be the most efficient if you have limited space. Once you've selected your system, gather the necessary components: a reservoir for the nutrient solution, pumps or air stones for oxygenation, tubing, and containers or channels for the plants.

An image showing hydroponics and aeroponics urban garden

Assembling and connecting the components is the next step. Start by setting up the reservoir and filling it with water. Add the appropriate nutrient solution, following the manufacturer's instructions for the correct concentration. Attach the pump or air stone to ensure the roots receive adequate oxygen. Connect the tubing to deliver the nutrient solution to the plants. In a drip system, for example, run the tubing from the pump to each plant, with drip emitters placed at the base of each plant. For an NFT system, set up the channels on a slight incline and connect the tubing to allow the nutrient solution to flow over the roots and back into the reservoir.

Preparing and maintaining the nutrient solutions is crucial for the success of your hydroponic or aeroponic garden. Start using a high-quality hydroponic nutrient mix specifically formulated to provide all the essential nutrients your plants need. Mix the solution according to the instructions and regularly check the concentration with a nutrient meter. Changing the nutrient solution every two weeks is essential to prevent the buildup of salts and other residues.

Monitoring and adjusting the pH and nutrient levels is also essential. Most plants thrive in a slightly acidic environment, with a pH between 5.5 and 6.5. Use a pH meter to regularly check the pH of your nutrient solution and adjust it as needed with pH up or down solutions. Keeping the pH within the optimal range ensures that your plants can absorb the nutrients effectively. Regularly check the nutrient levels with an EC (electrical conductivity) meter to ensure your plants receive the proper nutrients. Adjust the concentration as needed to maintain a balanced nutrient solution.

Hydroponics and aeroponics offer exciting possibilities for urban gardeners. They allow you to grow various plants in limited spaces with greater efficiency and higher yields. By understanding the principles of these soil-less gardening methods and carefully setting up and maintaining your systems, you can create a thriving urban garden that maximizes your resources and produces abundant, healthy plants.

9.3 Soil Microbiology: Understanding Beneficial Microorganisms

Soil microorganisms are pivotal in maintaining soil health and promoting plant growth. These tiny organisms are the backbone of soil fertility, helping to break down organic matter and release essential nutrients. One of their key functions is enhancing nutrient availability. Microorganisms decompose complex organic materials into simpler compounds that plants can readily absorb. This process, known as mineralization, is crucial for making nutrients like nitrogen, phosphorus, and sulfur available to plants. Without these microorganisms, many essential nutrients would remain locked in forms that plants cannot use.

Another critical role of soil microorganisms is promoting root health and growth. Certain bacteria, such as those in the genus Rhizobium, form symbiotic relationships with leguminous plants. These nitrogen-fixing bacteria convert atmospheric nitrogen into a form that plants can absorb, significantly boosting soil fertility. Mycorrhizal fungi, another group of beneficial microorganisms, form mutualistic associations with plant roots. These fungi extend the root system through their hyphae, increasing the plant's ability to absorb water and nutrients, particularly phosphorus. This symbiotic relationship enhances nutrient uptake and improves plant resilience to stress and disease.

Soil microorganisms also play a vital role in suppressing soil-borne diseases. Beneficial microbes can out-compete harmful pathogens for resources and space, reducing the incidence of diseases like root rot and damping-off. They produce antibiotics and other bioactive compounds that inhibit the growth of pathogens. Moreover, some beneficial microorganisms induce systemic resistance in plants, making them more resistant to various diseases. This natural form of pest and disease control reduces the need for chemical pesticides, promoting a healthier and more sustainable garden ecosystem.

Different beneficial microorganisms commonly found in healthy soil include bacteria and fungi. Nitrogen-fixing bacteria, such as Rhizobium and Azotobacter, are essential for converting atmospheric nitrogen into a form that plants can use. These bacteria form nodules on the roots of leguminous plants, where they fix nitrogen in exchange for carbohydrates from the plant. Decomposing bacteria, such as Bacillus and Pseudomonas, break down organic matter, releasing nutrients into the soil. Mycorrhizal fungi, including species from the genera Glomus and Rhizophagus, form mutualistic associations with plant roots. These fungi increase nutrient uptake, particularly phosphorus, and enhance plant resilience. Decomposer fungi, such as Trichoderma and Penicillium, break down complex organic materials, releasing nutrients and improving soil structure.

Promoting and maintaining a healthy population of beneficial microorganisms in the soil involves several strategies. One of the most effective methods is adding compost and organic matter to the soil. Compost provides a rich source of organic material that feeds beneficial microbes, enhancing their activity and diversity. Regularly incorporating compost into your garden soil can significantly improve microbial health. Avoiding chemical fertilizers and pesticides is another crucial practice. These chemicals can harm beneficial microorganisms and disrupt the soil ecosystem. Instead, use organic fertilizers and natural pest control methods that support microbial life. Microbial inoculants and biofertilizers can also be beneficial. These products contain live beneficial microorganisms that can boost soil health and plant growth. Applying microbial inoculants during planting or as a soil drench can introduce or enhance helpful microbial populations.

Monitoring and assessing the activity of soil microorganisms is essential for understanding soil health. Soil respiration tests measure microbial activity by determining the amount of carbon dioxide soil organisms produce. A high rate of respiration indicates a healthy, active microbial population. Microbial biomass assays measure the total microbial mass in the soil, offering insights into the overall microbial health. Observing plant health and growth can also serve as an indicator of soil microbial activity. Healthy, vigorous plants with robust root systems often indicate a thriving microbial community. Regularly monitoring these indicators can help you assess the effectiveness of your soil management practices and make necessary adjustments to support a healthy soil ecosystem.

9.4 Soil pH and Nutrient Management

Understanding soil pH is crucial for any urban gardener to cultivate healthy plants. Soil pH measures the acidity or alkalinity of soil on a scale from 0 to 14, with 7 being neutral. Values below 7 show acidity, while values above 7 indicate alkalinity. The pH level significantly impacts nutrient availability and uptake. In highly acidic soils (low pH), essential nutrients like phosphorus and magnesium become less available to plants. Conversely, iron and manganese may become less soluble in highly alkaline soils (high pH), leading to deficiencies. Optimal pH levels, usually between 6.0 and 7.0, ensure that nutrients remain accessible to plant roots, promoting robust growth and health.

Testing and adjusting soil pH is a straightforward process but requires careful attention. Start using pH testing kits or probes, readily available at garden centers or online. Collect soil samples from different garden areas to ensure an accurate reading. Mix the samples and follow the kit instructions, often involving mixing soil with distilled water and testing with a pH meter or test strip. If your soil is too acidic, adding lime can raise the pH. Lime, available in powdered or pellet form, should be evenly spread over the soil and then watered in. For best results, lime should be applied in the fall to allow time to adjust the soil pH by spring planting. Adding sulfur or peat moss can lower the pH if your soil is too alkaline. Sulfur is best applied in small amounts and is mixed thoroughly. Peat moss also lowers pH and adds organic matter, improving soil texture and water retention.

Effective nutrient management is essential for balanced and adequate plant nutrition. Regular soil testing helps identify nutrient levels, allowing you to tailor your fertilization practices. Use organic fertilizers and amendments, such as compost, worm castings, and bone meal, to provide a steady supply of nutrients. These organic materials improve soil structure and support beneficial microorganisms. Implementing crop rotation and cover cropping further enhances nutrient management. Crop rotation prevents the depletion of specific nutrients by alternating crops with different nutrient needs. Cover crops, such as clover or rye, enrich the soil by fixing nitrogen and adding organic matter. After the cover crops are tilled into the soil, they decompose, releasing nutrients for subsequent plantings.

Recognizing signs of nutrient deficiencies and toxicities is critical to addressing issues promptly. Nitrogen deficiency often manifests as the yellowing of older leaves, as nitrogen is mobile and moves to new growth, leaving older leaves deprived. Phosphorus deficiency can cause leaves to turn purple or dark green, with stunted growth and delayed maturity. In contrast, iron toxicity, often from overly acidic soils, leads to bronzing or stippling of the leaves, especially in younger foliage. These visual symptoms can guide you in diagnosing and correcting nutrient imbalances. Applying the appropriate organic amendments can quickly remedy deficiencies. For example, adding compost or blood meal can address nitrogen deficiencies, while bone meal can correct phosphorus shortages.

Maintaining an optimal soil pH and proper nutrient balance is a dynamic process that requires regular monitoring and adjustment. By understanding the concept of soil pH and its impact on nutrient availability, you can create an environment where plants thrive. Regular soil testing, organic fertilizers, and strategic practices like crop rotation and cover cropping ensure your soil remains fertile and productive. Recognizing and addressing signs of nutrient deficiencies and toxicities further supports plant health and growth. This comprehensive approach to soil pH and nutrient management is essential for any urban gardener seeking to maximize their garden's potential.

9.5 Biochar: Enhancing Soil Fertility in Urban Gardens

Biochar is a fascinating and powerful tool for enhancing soil fertility, especially in urban gardens where soil quality can be a limiting factor. Biochar is a carbon-rich material produced by heating organic matter, such as wood chips or crop residues, without oxygen through pyrolysis. This method transforms the organic material into a stable form of carbon that you can add to the soil. The benefits of biochar are numerous. It improves soil structure by creating a porous matrix that enhances aeration and water retention. It also helps to prevent soil compaction and allows roots to penetrate more easily.

Additionally, biochar increases nutrient retention, reducing the need for frequent fertilization. Its porous nature acts like a sponge, holding on to essential nutrients and releasing them slowly to plants. Furthermore, biochar enhances microbial activity. The pores in biochar provide a habitat for beneficial soil microorganisms, promoting a thriving soil ecosystem.

An image showing biochar use in an urban garden

Making and using biochar in your urban garden is a straightforward process that starts with selecting suitable organic waste materials. You can use wood chips, agricultural residues, or even certain types of manure. To produce biochar, you will need a pyrolysis setup, which can be as simple as a double-barrel retort system. Place the organic material in one barrel and heat it without oxygen, allowing it to carbonize without burning. This process typically takes a few hours, and safety precautions are essential, including wearing gloves and using a fire extinguisher. Once the biochar is produced and cooled, it can be integrated into your soil.

Integrating biochar into your garden soil involves mixing it with compost and other soil amendments. This combination enhances the overall fertility and structure of the soil. Begin by crushing the biochar into smaller pieces to increase its surface area, then mix it with compost at a ratio of about one part biochar to ten parts compost. Spread this mixture over your garden beds and work it into the top few inches of soil. Mix biochar directly into the potting soil for container gardens at a similar ratio. The application rates can vary depending on the soil conditions, but a general recommendation is to apply biochar at one to three pounds per square foot of the garden area.

The impact of biochar on soil microbiology is profound. Biochar increases microbial diversity and activity by providing a stable habitat for beneficial microorganisms. The porous structure of biochar offers shelter for microbes, protecting them from environmental stressors and helping them thrive. This enhanced microbial activity accelerates nutrient cycling and decomposition, making nutrients readily available to plants. Studies have shown that biochar can significantly increase the population of mycorrhizal fungi and other beneficial microbes, leading to healthier and more resilient plants. The increased microbial activity also helps suppress soil-borne diseases, as beneficial microbes outcompete harmful pathogens for resources and space.

Case studies and research on the use of biochar in urban gardens have demonstrated its long-term benefits. In community gardens, biochar has significantly improved soil fertility and plant growth. For example, a community garden in New York City experienced a marked increase in vegetable yields after incorporating biochar into its soil. The gardeners noted that their plants were healthier, with stronger root systems and more robust growth. Similarly, research on rooftop gardens has highlighted the long-term benefits of biochar. A study conducted in Berlin found that biochar improved soil structure and fertility over several growing seasons, leading to sustained plant growth and reduced need for chemical fertilizers. These case studies underscore the potential of biochar to transform urban gardening by creating healthier, more productive soil.

Incorporating biochar into your urban garden can enhance soil fertility, structure, and microbial activity. You can create a thriving garden environment that supports robust plant growth by producing biochar from organic waste materials and integrating it into your soil. The positive impact on soil microbiology further enhances nutrient cycling and disease suppression, leading to healthier plants and higher yields. Case studies and research demonstrate the long-term benefits of biochar, making it a valuable addition to any urban garden. As you explore the potential of biochar, you will find that this sustainable practice improves your garden and contributes to a healthier environment.

By focusing on advanced soil management techniques like no-till gardening and hydroponics, understanding soil microbiology, and using biochar, you can create a thriving urban garden that maximizes productivity while promoting sustainability. These methods offer practical solutions to urban gardeners' everyday challenges, enhancing soil health and plant growth. As you continue to explore these techniques, you'll discover new ways to optimize your garden and contribute to a more sustainable urban environment.

Chapter Ten

Designing and Maintaining Aesthetic Urban Gardens

One winter, I found myself staring at the bare, stark space that was in my backyard. The cold, gray environment mocks my vision of a lush, vibrant garden. Determined to bring life and beauty into this small urban space, I explored ways to design and maintain an aesthetically pleasing garden. This chapter is about transforming your limited space into an inviting, visually stunning garden that reflects your style and maximizes every available area.

10.1 Designing Small Garden Spaces: Layout and Aesthetics

Maximizing space efficiency is crucial for urban gardening. With limited space, every square inch must serve a purpose. One effective technique is vertical gardening, which allows you to grow plants upward rather than outward. You can create a lush, green wall that adds depth and beauty to your space using trellises, wall-mounted planters, and hanging baskets. Multi-level planters and tiered beds are also excellent for maximizing planting areas. By stacking plants vertically, you can grow various species in a compact footprint. Mirrors can create the illusion of more space by reflecting light and greenery, making your garden feel more extensive and more open.

Creating focal points in your garden design draws the eye and adds interest. A well-placed water feature like a small fountain or pond can serve as a calming centerpiece. The sound of trickling water adds a soothing element to the garden. Statement plants or trees can also act as focal points. Choose plants with unique shapes, colors, or textures to stand out. Garden sculptures and art pieces provide an artistic touch, reflecting your personality and style. These elements can be both functional and decorative, enhancing the overall aesthetic of your garden.

Color schemes and plant selection are vital for a cohesive look. Harmonizing colors creates a uniform and pleasing visual experience. Consider using a color wheel to select complementary colors that work well together. Seasonal color changes with flowering plants like spring tulips or autumn chrysanthemums can keep your garden vibrant year-round. The presence of contrasting foliage in the surroundings heightens visual interest. Plants with different leaf shapes, sizes, and colors can create a dynamic and engaging garden environment. Mixing greens with red, purple, or yellow splashes can make your garden visually appealing.

Pathways and access are functional and contribute to the garden's aesthetics. Stepping stones or gravel paths can guide visitors through the garden, creating a sense of exploration. Winding paths add a whimsical touch and make the space feel larger. Ensure accessibility for all users by designing paths that are wide enough and smooth enough for easy navigation. This design is essential for sharing the garden with friends, family, or community members. Thoughtfully designed pathways can enhance both the beauty and usability of your garden.

Garden Design Checklist

- **Vertical Gardening:** Utilize trellises, wall planters, and hanging baskets.

- **Multi-Level Planters:** Incorporate tiered beds to maximize planting space.

- **Focal Points:** Add water features, statement plants, or sculptures.

- **Color Schemes:** Harmonize colors and use seasonal plants for variety.

- **Pathways:** Create functional and aesthetically pleasing paths with stepping stones or gravel.

Enhancing your garden with these design principles can transform even the smallest urban space into a beautiful, functional haven. Whether you grow vegetables or flowers or enjoy the greenery, thoughtful design can make your garden a valid extension of your home.

10.2 Elevating Small Gardens into Serene Sanctuaries

Creating a sense of privacy in your small garden can transform it into a personal retreat. One of the best ways to achieve this is by using tall plants or trellises to create natural screens. Bamboo, tall grasses, or even fast-growing shrubs can provide an immediate sense of the enclosure. Trellises can offer privacy and beauty when adorned with climbing plants like jasmine or clematis; installing decorative fences or walls is another effective method. You can customize these to match your garden's aesthetic while providing a solid barrier. Adding pergolas or arbors with climbing plants creates shade and adds a vertical dimension, making your garden feel more secluded and intimate.

Water elements can be calming, turning your garden into a serene sanctuary. Installing small fountains or birdbaths can introduce the soothing sound of trickling water, which can mask urban noise and create a peaceful atmosphere. Container water gardens are another fantastic option. These can be simple to set up and maintain, using large pots or barrels filled with water plants like lilies and lotus. Consider creating DIY water features with recycled materials for a more personalized touch. You can repurpose old pots, ceramic jars, or even wine barrels into unique water elements that add charm and tranquility to your garden.

An image showing the aesthetic value of an urban garden

Using lighting to enhance the ambiance of your garden can extend its usability into the evening hours. Solar-powered Garden lights are an eco-friendly option that requires minimal maintenance. You can place them along pathways, around seating areas, or among plants to create a soft, inviting glow. String lights can add a cozy ambiance, perfect for evening gatherings or quiet outdoor nights. Drape them over pergolas, along fences, or even in trees to create a magical atmosphere. Using spotlights, you can make a more dramatic and vibrant nighttime look for your garden by emphasizing key features like plants, water elements, or art.

Including comfortable seating areas invite relaxation and contemplation. Choose comfortable and weather-resistant furniture that suits the style of your garden. Options like cushioned benches, lounge chairs, or hammocks can provide a cozy place to unwind. Creating shaded seating with umbrellas or canopies ensures you can enjoy your garden even on hot, sunny days. Integrating seating into garden structures like benches or swings adds functionality and charm. A swing nestled among lush greenery or a bench under a flowering arbor can become a favorite spot for reading, meditating, or simply enjoying the sights and sounds of your garden.

Focusing on privacy, water elements, lighting, and comfortable seating can transform your small garden into a serene sanctuary. These elements work together to create a space that is not only beautiful but also a haven of peace and relaxation.

10.3 Incorporating Art and Decor in Urban Gardens

Selecting garden art can bring personality and charm to your urban garden. When choosing pieces, consider what complements the overall design and reflects your style. Sculptures and statuary can add a touch of elegance or whimsy, depending on the theme you choose. Garden gnomes and other whimsical figures can bring a playful, enchanting feel, making your garden a delightful space for children and adults. Artistic planters and containers, like those made from unique materials or with intricate designs, can serve as focal points while housing your favorite plants.

For a truly personal touch, consider DIY garden decor projects. Painted rocks and stepping stones can be a fun and creative way to add color and interest to your garden paths. Each stone can be a small canvas for your artistic expression, whether you paint flowers, animals, or abstract designs. Handmade wind chimes and mobiles, crafted from recycled materials like old silverware or seashells, can bring gentle, soothing sounds to your garden. Upcycled garden signs and labels, made from repurposed wood or metal, can add a rustic charm while helping you keep track of your plants.

An image showing decorative trellises in the urban garden and the charm of a cityscape oasis

Incorporating functional decor serves a practical purpose while enhancing your garden's aesthetic. Decorative trellises and plant supports can guide climbing plants and add vertical interest. Artistic rain chains and water collectors can turn a rainy day into a visual and auditory delight while conserving water for your garden. Ornamental birdhouses and feeders attract birds, adding life and movement to your garden. They also serve as charming decorative elements that you can customize to match your style.

Changing decor seasonally or creating themed garden areas can keep your garden fresh and exciting throughout the year. Seasonal wreaths and garlands from natural materials like pinecones, leaves, and flowers can celebrate the changing seasons. Holiday-specific decorations, such as twinkling lights for winter or colorful lanterns for summer, can add to a festive spirit. Themed garden areas, like fairy gardens with tiny houses and figurines or Zen gardens with raked sand and stone arrangements, can provide unique spaces within your garden for different moods and activities.

Incorporating art and decor in your urban garden blends functionality with personal expression. Each element, from sculptures to seasonal decorations, adds depth and character, making your garden a space for plants and reflecting your creativity and style.

10.4 Community Gardening: Building Connections and Sharing Knowledge

Participating in a community garden offers many benefits beyond the garden itself. Gardeners share resources and knowledge, fostering a strong sense of community and belonging. This collaboration promotes sustainability and helps bolster local food production. In cities with limited space, community gardens provide a green oasis where neighbors can connect, exchange tips, and work towards common goals. The social aspect of community gardening is significant—it brings people together, creating bonds that strengthen the fabric of the community.

Starting a community garden involves a few key steps. First, securing a location is crucial. Look for underutilized spaces such as vacant lots, rooftops, or school grounds. Once you identify a spot, you must obtain the necessary permits from local authorities. Organizing a group of interested participants is the next step. Reach out to neighbors, regional organizations, and schools to gather a diverse group of gardeners. Planning and designing the garden layout comes next. Consider the community needs, including accessibility, types of plants, and shared spaces like seating areas or tool sheds. Sketch out the design and designate specific plots for different gardeners or groups.

Managing a community garden requires clear communication and organization. Establishing garden rules and guidelines helps ensure that everyone understands their responsibilities. These rules cover watering schedules, composting practices, and plot maintenance. Dividing responsibilities and tasks among members can make the garden more manageable. Create committees for different aspects of the garden, such as maintenance, events, and education. Hosting regular meetings and events keeps everyone engaged and informed. These gatherings can be great for sharing updates, addressing issues, and celebrating successes.

Community gardens also serve as valuable educational spaces for both children and adults. Organizing workshops and classes can provide hands-on learning opportunities. Topics might include composting, organic pest control, or seasonal planting techniques. Creating educational signage and displays around the garden can offer ongoing learning. These signs might explain different plant species, beneficial insects, or sustainable gardening practices. Partnering with local schools and organizations can further enhance the educational impact. Schools might use the garden for science classes, while regional organizations could host workshops or volunteer days.

Garden Start-Up Checklist

- **Secure Location:** Identify and obtain permits for the garden site.

- **Organize Participants:** Gather a diverse group of interested gardeners.

- **Plan Layout:** Design a layout that meets the community's needs.

- **Establish Rules:** Create clear guidelines for garden maintenance and responsibilities.

- **Host Meetings:** Hold regular gatherings to keep everyone engaged.

- Community gardening is a powerful way to build connections and share knowledge.

Communities can work together to create vibrant, productive gardens that provide fresh food, educational opportunities, and a sense of belonging.

Chapter Eleven

Success Stories: Inspirational Urban Gardens from Around the World

U rban gardening has sprouted in cities across the globe, transforming urban landscapes into lush, productive havens. In New York City, rooftop gardens have become an oasis amidst the concrete jungle. One such garden, perched atop a Brooklyn apartment building, has turned an otherwise unused space into a thriving ecosystem. The gardener maximized the limited space using vertical gardening systems and recycled materials. The vertical planters, filled with herbs and vegetables, cascade down the walls, creating a green tapestry that offers beauty and sustenance. Hydroponic setups further enhance productivity, allowing for year-round cultivation with no soil.

In Tokyo, balcony gardens have become a testament to innovation and resilience. One gardener, working with a mere five square meters, transformed her balcony into a vibrant garden using a mix of vertical gardening and container planting. By employing multi-level planters and tiered beds, she grew an impressive variety of plants, including tomatoes, peppers, and lettuce. Mirrors strategically placed along the walls created an illusion of a larger space, enhancing the garden's visual appeal. The gardener's dedication to eco-friendly practices, such as composting kitchen scraps and using organic fertilizers, ensured a sustainable and fertile growing environment.

Berlin hosts many community gardens that serve as green sanctuaries amidst urban sprawls. One standout project turned an abandoned lot into a thriving community garden, fostering connections among neighbors and promoting local food production. The gardeners created an inclusive space where everyone could contribute by using raised beds made from upcycled materials. Innovative design solutions, such as rainwater harvesting systems and solar-powered irrigation, addressed common challenges like water usage and conservation. The garden's success hinges on the community's collective effort, with each member bringing unique skills and knowledge.

These gardens have not only overcome physical challenges but also inspired personal transformations. The rooftop gardener in Brooklyn, motivated by a desire to reconnect with nature, found solace and purpose in cultivating his urban oasis. Despite initial setbacks, like soil contamination and limited sunlight, he persevered by implementing organic soil remediation techniques and using reflective surfaces to enhance light exposure. The Tokyo gardener, driven by a passion for sustainability, faced challenges such as pests and space constraints. Her innovative use of vertical gardening and natural pest control methods turned obstacles into opportunities for growth. In Berlin, the community garden became a beacon of hope for residents seeking a sense of belonging and purpose. The gardeners faced hurdles like securing funding and navigating regulations, but their shared vision and determination led to a flourishing green space that benefits all.

These personal stories highlight urban gardeners' motivations, challenges, and triumphs worldwide. Whether dealing with space constraints, soil issues, or water management, these gardeners have shown that creativity, resilience, and dedication can transform even the most unlikely spaces into thriving urban gardens. Their experiences offer valuable lessons and inspiration for anyone looking to cultivate their green oasis in the city.

11.1 Mindfulness in Gardening: Health and Wellness Benefits

Gardening offers profound mental health benefits, providing a sanctuary to escape daily stressors and immerse yourself in nature. The calming effect of being surrounded by plants, with their gentle rustling and vibrant colors, creates a space where your mind can find peace. While gardening, mindfulness practices, such as focusing on the texture of the soil or the scent of herbs, help you stay present. Each moment spent tending to your garden can become a meditation, allowing you to clear your mind and reduce anxiety. By dedicating time to your garden, you create a routine that fosters mental well-being, offering a respite from the hustle and bustle of urban life.

The physical health benefits of gardening are equally significant. Engaging in light to moderate physical activity, such as digging, planting, and weeding, helps improve your overall fitness. These activities enhance flexibility and strength as you bend, stretch, and lift. Gardening also provides an opportunity to boost your vitamin D levels from sunlight exposure, essential for bone health and immune function. Spending time outdoors, surrounded by fresh air, contributes to your physical well-being, making gardening a holistic activity that nurtures both body and mind.

Creating a mindful gardening routine can further enhance the benefits you reap from this practice. Start by focusing on the present moment, letting go of distractions, and immersing yourself in the task. Engage all your senses during gardening activities, feel the soil, listen to the birds, and observe the growth of your plants. You are practicing gratitude and reflection while gardening can deepen your connection to nature. Take a moment to appreciate the life you're nurturing and reflect on your gardening journey, fostering a sense of accomplishment and peace.

Therapeutic gardening practices can be beneficial for specific health conditions. Horticultural therapy programs, often led by trained therapists, use gardening activities to improve physical and mental health. These programs can be tailored to individuals with various needs, offering a supportive environment for healing. Sensory gardens, designed with plants stimulating the senses, benefit individuals with disabilities. These gardens can include fragrant herbs, colorful flowers, and textured leaves, providing a rich sensory experience. Adaptive gardening tools and techniques ensure that gardening remains accessible to everyone, regardless of physical limitations. Tools with ergonomic handles, raised beds, and container gardening can make gardening more manageable and enjoyable for individuals with mobility issues.

Incorporating mindfulness into your gardening routine transforms it into a therapeutic practice that benefits your mental and physical health. Whether seeking a peaceful escape, a way to stay active, or therapeutic support, gardening offers a multifaceted approach to wellness. Embrace the opportunity to connect with nature, nurture your plants, and promote your well-being.

11.2 Keeping Your Garden Thriving: Seasonal Tips and Tricks

Spring is a time of renewal and growth, making it the perfect season to prepare your garden for the year ahead. Start by cleaning up the winter debris. Remove dead leaves, branches, and remnants of last year's plants. This practice makes your garden look tidy and prevents pests and diseases from taking hold. Next, focus on preparing the soil. Turn it over, break up clumps, and add compost to enrich it with nutrients. Plant cool-season crops like lettuce, spinach, and peas as they thrive in the mild temperatures of early spring. Remember to set up your irrigation systems to ensure consistent watering. Drip irrigation is adequate for precise moisture control.

When summer arrives, the heat can be a blessing and a challenge for your garden. Mulching becomes crucial. Apply a thick layer of organic mulch around your plants to retain soil moisture and keep the roots cool. Mulch also helps suppress weeds. Providing shade for sensitive plants is another important task. Use shade cloths or strategically place taller plants to shield more delicate species from the intense sun. Summer is also a prime time for pests and diseases. Regularly inspect your plants for signs of trouble, such as discolored leaves or unusual spots. Organic pest control methods, like neem oil or insecticidal soap, can help manage these issues without harming the environment.

Fall is a season of transition, and your garden requires specific care to stay productive. Start by harvesting late-season crops like pumpkins, squash, and root vegetables. Add compost and organic matter to the soil to replenish nutrients as you clear out spent plants. This season is also the time to plant cover crops, such as clover or rye, which will protect the soil over the winter and improve its structure. Bulbs for spring flowers should be planted now, giving them time to establish roots before the ground freezes. These tasks prepare your garden for a restful winter and a vibrant start in the spring.

Winter gardening may seem daunting, but your garden can endure the cold months with the right strategies. Protecting plants from frost and cold is essential. Use row covers, cloches, or old blankets to shield them from freezing temperatures. Pruning perennials during winter dormancy helps maintain their shape and promotes healthy growth in the spring. Mulching perennials and other vulnerable plants insulate their roots, keeping them safe from temperature fluctuations. Winter is also the perfect time to plan and order seeds for spring. Take this opportunity to assess what worked well in your garden and what didn't and make plans for new plants and designs. This preparation ensures that your garden will be ready to burst into life when the weather warms up again.

11.3 Future Trends in Urban Gardening: Innovations and Predictions

Technological innovations are revolutionizing urban gardening, making it more efficient and accessible. Smart gardening devices and apps allow you to monitor soil moisture, light levels, and plant health from your phone. These tools provide real-time data and helpful tips, ensuring your plants get nutrients. Automated irrigation and fertilization systems take the guesswork out of maintaining your garden. These systems optimize plant growth and conserve resources by delivering precise amounts of water and nutrients. Advanced hydroponics and vertical farming techniques are also gaining popularity.

These methods allow for high-density planting in small spaces, using nutrient-rich water or vertical structures to grow plants without soil. They are instrumental in urban areas where ground space is limited.

Emerging sustainable practices are setting the stage for a greener future in urban gardening. Zero-waste gardening aims to eliminate waste by composting organic matter and reusing materials. This practice reduces your environmental footprint and enriches your soil. Climate-resilient gardening techniques are becoming essential as weather patterns become more unpredictable. Using drought-resistant plants, rainwater harvesting, and soil conservation methods can help your garden thrive despite changing conditions. Eco-friendly garden products and materials, such as biodegradable pots and organic fertilizers, are increasingly available. These products support sustainable gardening practices and reduce the impact on the environment.

Social trends are also shaping the future of urban gardening. Community gardens are becoming more popular, fostering connections and collaboration among neighbors. These shared spaces provide fresh produce and create a sense of community. There is a growing interest in edible landscapes, where ornamental plants are replaced with fruit trees, vegetables, and herbs. This trend promotes self-sufficiency and local food production. Urban gardening is also being recognized as a tool for social change. Gardens can beautify neglected areas, provide food security, and create green spaces that improve mental and physical health.

Looking ahead, the future of urban gardening holds exciting possibilities. Urban agriculture will expand, integrating more seamlessly into city planning. Rooftops, balconies, and even walls will become productive growing spaces. There will be an increased focus on biodiversity and native plants, supporting local ecosystems and attracting beneficial insects. As technology advances, we can expect innovative solutions that make gardening more accessible and sustainable. Whether it's through smart devices, sustainable practices, or community initiatives, urban gardening is poised to play a significant role in the future of our cities.

Conclusion

As we reach the end of this journey together, let's take a moment to reflect on our journey. We began by exploring the fundamentals of soil science, understanding its vital role in urban gardening, and learning how to manage soil in containers, raised beds,and limited spaces. We saw how healthy soil leads to thriving plants and how you can transform even the smallest urban area into a lush garden.

You learned about the importance of soil testing and the various techniques to amend and improve soil quality from the early chapters. You saw how adding compost, organic matter, and soil conditioners can make a significant difference. We delved into composting and vermiculture, discovering how to turn kitchen waste into valuable soil nutrients.

We then explored container gardening, discussing selecting suitable containers, creating the perfect soil mix, and implementing effective watering techniques. Vertical gardening emerged as an intelligent solution for maximizing space, and we also examined the best vegetables and herbs for container gardening.

Raised bed gardening brought strategies from designing and building raised beds to soil preparation and planting techniques. You learned about efficient irrigation solutions and the importance of seasonal maintenance and crop rotation to keep your garden productive year-round.

Organic pest and disease management provides eco-friendly ways to protect your plants. We explored identifying common pests, encouraging beneficial insects, and using homemade organic sprays. We also touched on the importance of soil health in disease prevention and the benefits of companion planting.

The advanced soil management techniques introduced you to no-till gardening, soil-less systems like hydroponics and aeroponics, and the role of soil microbiology. We also discussed the importance of soil pH and nutrient management and how biochar can enhance soil fertility.

We focused on designing and maintaining aesthetic urban gardens as we approached the end. You learned about maximizing small spaces, creating serene sanctuaries, and incorporating art and decor. We also highlighted the value of community gardening and mindfulness in gardening, showcasing this rewarding hobby's health and wellness benefits.

Throughout this book, several important points stand out:

1. The health of your soil is paramount. With good soil, plants can thrive.

2. You can overcome space limitations with creativity and smart techniques like vertical and container gardening.

3. Sustainable practices like composting, organic pest control, and water conservation benefit your garden and the environment.

4. The joy of gardening extends beyond just growing plants.

It fosters community, promotes mental and physical well-being, and connects you with nature.

Reflecting on my journey, I remember the challenges of starting an urban garden in a small space. There were moments of frustration and doubt, but the rewards outweighed the struggles. Gardening taught me patience, resilience, and the incredible satisfaction of seeing seeds transform into thriving plants. I hope you have found inspiration and practical guidance on these pages. Remember, every small step you take towards better soil management and sustainable gardening makes a difference.

Now, I encourage you to put what you have learned into action. Start by testing your soil, amending it with compost, and selecting suitable containers or building raised beds. Experiment with different plants, try vertical gardening, be bold, and make mistakes. Gardening is a continuous learning process, and each season brings new lessons.

For those who want to delve deeper into specific topics, I recommend the following resources:

1. "The Urban Gardener" by Matt James

2. "Teaming with Microbes" by Jeff Lowenfels and Wayne Lewis

3. "The Vegetable Gardener's Bible" by Edward C. Smith

4. Local extension services and community garden organizations

5. Online forums and gardening communities for shared experiences and advice

As we conclude, thank you for joining me on this journey. Your dedication to urban gardening and sustainable practices is commendable. Together, we can create greener, healthier urban spaces. Remember, every plant you grow contributes to a larger vision of sustainability and self-sufficiency.

Thank you for allowing me to share in your gardening adventure. I wish you many seasons of bountiful harvests, vibrant flowers, and the deep satisfaction of nurturing your urban garden. Happy gardening!

Chapter Twelve

End-of-Book Review Page

Title: **Soil Science for Urban Gardeners the** *ABCs of Growing Bloomy Plants in Containers, Raised Beds, and Limited Spaces in All Seasons for Healthy Living*

Dear Reader,

Congratulations on completing *Soil Science for Urban Gardeners*! Whether you are growing lush flowers in a container garden or cultivating vibrant vegetables in raised beds, you have just taken a big step toward creating a greener, healthier world. We hope this book has inspired you to maximize small spaces, grow your fresh produce, and embrace eco-friendly practices in your gardening journey.

Why Your Opinion Matters: your honest review of this book can make a difference for urban gardeners worldwide who, like you, are eager to:

- **Maximize Space Efficiency:** Unlock the secrets to gardening in small or unconventional spaces.

- **Embrace Sustainability:** Reduce their environmental impact while growing fresh, organic produce.

- **Learn and Educate:** Expand their knowledge of soil science and sustainable agriculture.

- **Enhance Wellness:** Enjoy the physical and mental health benefits of gardening year-round.

- **Build Community:** Connect with others through shared gardening efforts and sustainable living practices.

By sharing your feedback on Amazon, you are helping urban gardeners across North America, Europe, Africa, the Middle East, and Australia discover this valuable resource. This book is for men and women, from beginners to experienced growers, who are all passionate about sustainability, self-sufficiency, and creating thriving gardens.

How to Share Your Passion. Click the link below to leave your review on Amazon. Your review will help other gardeners find the tools they need to succeed and keep the spirit of *Soil Science for Urban Gardeners* alive for future readers.

>>> **Click here to leave your review on Amazon** <<<

Thank you for helping to cultivate a global community of urban gardeners!

Warm regards, *The Soil Science for Urban Gardeners Team*

P.S. Gardening is a journey, and every step counts. Your review is vital to spreading the knowledge and passion for sustainable gardening. Let us grow together!

Chapter Thirteen

References

1. SARE. (n.d.). *Soils for urban farms, gardens, and green spaces.* Retrieved from https://www.sare.org/publications/building-soils-for-better-crops/soils-for-urban-farms-gardens-and-green-spaces/

2. Frontiers in Environmental Science. (2018). *Urban soil quality assessment—A comprehensive case study.* Retrieved from https://www.frontiersin.org/journals/environmental-science/articles/10.3389/fenvs.2018.00136/full

3. Soil Science Society of America. (n.d.). *Soil contaminants.* Retrieved from https://www.soils.org/about-soils/contaminants#:~:text=Common%20contaminants%20in%20urban%20soils

4. Each Green Corner. (2021, April 3). *Choosing the right*

plants for your microclimate. Retrieved from https://www.eachgreencorner.org/2021/04/03/choosing-the-right-plants-for-your-microclimate/

5. Utah State University Extension. (n.d.). *Urban garden soils: Testing and management.* Retrieved from https://digitalcommons.usu.edu/extension_curall/2116/

6. Crop Nutrition. (n.d.). *Five benefits of soil organic matter.* Retrieved from https://www.cropnutrition.com/resource-library/five-benefits-of-soil-organic-matter/

7. Piedmont Master Gardeners. (n.d.). *Composting options for small, indoor, and restricted spaces.* Retrieved from https://piedmontmastergardeners.org/article/composting-options-for-small-indoor-and-restricted-spaces/

8. Cornell Composting. (n.d.). *Six easy steps to setting up a worm bin.* Retrieved from https://compost.css.cornell.edu/worms/steps.html

9. Naples Botanical Garden. (n.d.). *Container gardening - DIY potting mix.* Retrieved from https://www.naplesgarden.org/container-gardening-diy-potting-mix/

10. Colby Digs Soil. (2014, April 5). *Container gar-*

den water-saving tips from a soil scientist. Retrieved from https://colbydigssoil.com/2014/04/05/container-garden-water-saving-tips-from-a-soil-scientist/

11. Bungalow. (n.d.). *Urban gardening 101: Vegetables to grow in pots.* Retrieved from https://bungalow.com/articles/how-to-create-an-urban-garden-with-vegetables-in-pots

12. Green.org. (2024, January 30). *The role of vertical gardens in sustainable urban development.* Retrieved from https://green.org/2024/01/30/the-role-of-vertical-gardens-in-sustainable-urban-development/

13. University of Maryland Extension. (n.d.). *The safety of materials used for building raised beds.* Retrieved from https://extension.umd.edu/resource/safety-materials-used-building-raised-beds

14. Kellogg Garden. (n.d.). *Layering soil for an inexpensive raised garden bed.* Retrieved from https://kellogggarden.com/blog/raised-beds/layering-soil-for-an-inexpensive-raised-garden-bed/

15. Farmer's Almanac. (n.d.). *Companion planting guide.* Retrieved from https://www.farmersalmanac.com/companion-planting-guide

16. Homestead and Chill. (n.d.). *How to install drip irrigation in raised garden beds.* Retrieved from https://homesteadand-chill.com/install-drip-irrigation-raised-beds/

17. Lobotany. (n.d.). *Identifying urban garden pests.* Retrieved from https://lobotany.com/identifying-urban-garden-pests/

18. Xerces Society. (n.d.). *Beneficial insects for natural pest control: Foliage scouting.* Retrieved from https://xerces.org/publications/scouting-guides/beneficial-insects-for-natural-pest-control-foliage-scouting

19. The Prairie Homestead. (2015, July). *Organic pest control garden spray recipe.* Retrieved from https://www.theprairiehomestead.com/2015/07/organic-pest-control-garden-spray.html

20. West Virginia University Extension. (n.d.). *Companion planting.* Retrieved from https://extension.wvu.edu/lawn-gardening-pests/gardening/garden-management/companion-planting

21. The Old Farmer's Almanac. (n.d.). *How to build a cold frame.* Retrieved from https://www.al-

manac.com/how-build-cold-frame

22. Better Homes & Gardens. (n.d.). *The 11 best grow lights, based on testing.* Retrieved from https://www.bhg.com/best-grow-lights-7255086

23. Penn State Extension. (n.d.). *Cool-season vs. warm-season vegetables.* Retrieved from https://extension.psu.edu/cool-season-vs-warm-season-vegetables

24. SARE. (n.d.). *10 ways cover crops enhance soil health.* Retrieved from https://www.sare.org/publications/cover-crops-ecosystem-services/10-ways-cover-crops-enhance-soil-health/

25. City Farmer News. (n.d.). *Urban permaculture garden in San Francisco grows thousands of pounds of food.* Retrieved from https://cityfarmer.info/urban-permaculture-garden-in-san-francisco-grows-thousands-of-pounds-of-food/

26. Santa Clara Valley Water District. (n.d.). *Rainwater harvesting systems.* Retrieved from https://www.valleywater.org/saving-water/outdoor-conservation/rainwater-harvesting-systems

27. Sustain My Craft Habit. (n.d.). *45 upcycling ideas*

for the garden. Retrieved from https://sustainmy-crafthabit.com/diy-upcycling-projects-for-the-garden/

28. Lyngso Garden Materials. (n.d.). *The advantages of organic fertilizers for your garden plants.* Retrieved from https://www.lyngsogarden.com/community-resources/using-organic-fertilizers-in-your-gardens/

29. Epic Gardening. (n.d.). *11 benefits of no-till gardening.* Retrieved from https://www.epicgardening.com/benefits-of-no-till-gardening/

30. Sensorex. (n.d.). *6 types of hydroponic systems explained.* Retrieved from https://sensorex.com/hydroponic-systems-explained/

31. NCBI. (n.d.). *The role of soil microorganisms in plant mineral nutrition.* Retrieved from https://www.ncbi.nlm.nih.gov/pmc/articles/PMC5610682/

32. Oklahoma State University Extension. (n.d.). *Preparation of biochar for use as a soil amendment.* Retrieved from https://extension.okstate.edu/fact-sheets/preparation-of-biochar-for-use-as-a-soil-amendment.html

33. Adams Bailey Company. (n.d.). *How to design a garden in limited space.* Retrieved from https://www.adamsbailey.com/about-us/blogs/gardening-in-urban-areas-how-to-design-a-garden-in-limited-space

34. Lawn Love. (n.d.). *Best plants for vertical gardens.* Retrieved from https://lawnlove.com/blog/best-plants-vertical-gardens/

35. Landscape East & West. (2012, March). *Landscape design strategy: How to create a focal point.* Retrieved from https://www.landscapeeast.com/blog/creating-a-focal-point-2012-03

36. World Bank. (n.d.). *Urban agriculture: Findings from four city case studies.* Retrieved from https://www.worldbank.org/en/topic/urbandevelopment/publication/urban-agriculture-four-city-case-studies